American Automobile
PAINT CODE
INTERCHANGE MANUAL
1945–1995

Peter C. Sessler

Motorbooks International
Publishers & Wholesalers ®

Special thanks to PPG Industries, Inc, Du Pont and Susan Bowden.

First published in 1995 by Motorbooks International Publishers & Wholesalers, PO Box 2, 729 Prospect Avenue, Osceola, WI 54020 USA

© Peter C. Sessler, 1995

Motorbooks International books are also available at discounts in bulk quantity for industrial or sales-promotional use. For details write to Special Sales Manager at the Publisher's address

Library of Congress Cataloging-in-Publication Data

Sessler, Peter C.
American automobile paint code interchange manual, 1945-1995/ Peter C. Sessler.
    p.   cm.
Includes index.
ISBN 0-87938-977-X
1. Automobiles--Painting--Standards--United States. 2. Automobiles--United States--Serial numbers. I. Title
TL255.2.S47   1995
629.26--dc20

On the front cover: 1969 Z28 Camaro. *Mike Mueller*

Printed and bound in the United States of America

CONTENTS

INTRODUCTION

American Automobile Paint Code Interchange Manual, 1941-1995 lists the paint codes used on passenger cars produced by General Motors, Ford Motor Co., Chrysler Corporation and American Motors Corporation from 1941 to 1995. The paint codes, along with the colors they represent are listed yearly. Also included are the paint codes used by a major paint manufacturer, PPG Industries (Ditzler). Due to space limitations, only PPG's codes are used, however, this should not be construed as an endorsement for PPG. The paints produced by other paint manufacturers such as Du Pont, Sikkens, Martin-Williams. Acme, BASF/R-M, Glasurit etc. are also of excellent quality. The listing of an aftermarket paint manufacturer's paint codes makes it easier to pin down and identify the correct color code for a particular car.

Paints consist of pigments, resins and solvents. Pigments provide color and also include the chemicals which make the paint what it is- its resistance to corrosion, its reflective properties and surface characteristics. Resins make it possible for the pigment to remain in suspension, along with any other material added to the paint, such as metalflakes. Solvents are used to facilitate the application, thinning, hardening, and drying of the paint.

Briefly, there are two types of paints used on automobiles—lacquers and enamels. Lacquer paints dry and harden by having their solvents evaporate into the air. Generally, lacquers, once buffed out, can provide a deeper shine but they aren't as durable or as stable as an enamel. Acrylic lacquers are more durable because an acrylic agent—a synthetic plastic hardener— is mixed in the paint's resin package.

Enamels dry and harden by the reaction of the chemicals used in their makeup rather than the evaporation of solvents. Enamel paint also does not need to be buffed out. In urethane paint, which is an enamel type paint, an additional chemical agent is used with the result being a much more durable, chemical resistant paint that rivals the gloss of a lacquer.

Like the cars they are used on, paints themselves are continually being improved.

They are better described now as "coating systems" consisting of fillers, etchers, primers, surfacers, reducers, hardeners, basecoats, metallic basecoats and gloss clears. If you want to learn more about paints and how to paint your car, there are several good books on the subject available from Classic Motorbooks. You can spend an amazing amount of money on a "coating system" for your car!

Usually, the paint codes listed in the book are found on a an identification plate located somewhere on the car. This can be on the door(s), under the hood, glove box—it all depends where the manufacturer placed the plate. To find the exact location of the i.d. plate on your car(if you can't readily locate it), refer to a factory shop manual or one of the other books in the Red Book Series. Although the codes listed here are correct, be aware that car manufacturers have used and continue to use colors that aren't normally listed. For example, if you own a Chrysler product that has the number 999 on the ID plate instead of a color code, this means that your car has a special order

color. Ford, on the other hand, simply left the color code area blank on the ID plate when a car was painted a special order color.

You'll note that from 1979-on, the codes and colors listed for each manufacturer's divisions are grouped together. Unlike previous years, the practice of giving the same color a different name stopped. For example, Chevrolet used a Tuxedo Black in 1973 while Oldsmobile's version, which had the same code, was called Ebony Black. From 1979-on, it was just Black.

Although every effort has been made to make sure that the information contained in the American Car Paint Code Interchange Catalog, 1941-95, is correct, I cannot assume any responsibility for any loss arising from the use of this book.

Peter C. Sessler
Milford, PA
December 1994

Chapter 1   AMERICAN MOTORS, 1957-1988

COLOR CODE	COLOR	DUPONT CODE
1957 RAMBLER		
P-67	Bermuda Green	2103
P-72	Frost White	2382
P-82	Pacific Blue	2383-H
P-84	Glacier Blue	2509
P-85	Lagoon Blue	2510
P-86	Plum	2533-H
P-87	Berkshire Green	2511
P-88	Oregon Green	2513-H
P-89	Avocado	2517
P-90	Mardi Gras Red	2514-H
P-92	Mojave Yellow	2515
P-93	Sierra Peach	2516
P-92	Cinnamon Bronze	2512-H
P-95	Gotham Gray	2518-H
1958 RAMBLER		
P-2	Kimberly Blue	2772
P-3	Saranac Green	2773
P-4	Alamo Beige	2774
P-5	Autumn Yellow	2775
P-6	Georgian Rose	2776
P-7	Mariner Turquoise	2777
P-72	Frost White	2382
P-94	Cinnamon Bronze	2512-H
P-90	Mardi Gras Red	2514-H
P-92	Mojave Yellow	2515
P-95	Gotham Gray	2518-H
P-97	Brentwood Green	2778
P-98	Lakelshore Blue	2779
P-99	Frontenac Gray	2780
1959 RAMBLER		
P-4	Alamo Beige	2774
P-5	Autumn Yellow	2775
P-8	Chatsworth Green	2970
P-9	Pineridge Green	2973
P-10	Placid Blue	2969
P-11	Nocturnc Blue	2974-H
P-12	Alladin Gray	2977
P-13	Oriental Red	2971-H
P-14	Carmel Copper	2915
P-15	Aqua Mist	2978
P-16	Cotillion Mauve	2972
P-17	Hibiscus Rose	2976

COLOR CODE	COLOR	PPG CODE
P-72	Frost White	2382
P-99	Frontenac Gray	2780
1960 RAMBLER		**PPG CODE**
P-1	Classic Black	9000
P-4	Alamo Beige	21491
P-5	Autumn Yellow	21492
P-8	Chatsworth Green Light	42463
P-10	Placid Blue Light	11985
P-13	Oriental Red	70946
P-15	Aqua Mist Poly	42462
P-18	Westchester Green Poly Med.	42648
P-19	Sovereign Blue Poly Med.	12220
P-20	Dartmouth Gray Light	31987
P-21	Harvard Gray Poly Med.	31988
P-23	Echo Green Poly	42692
P-24	Auburn Red Poly	71101
P-25	Festival Rose	71102
P-72	Frost White	8108
P-94	Cinnamon Bronze Poly	21303
1961 RAMBLER		
P-1	Classic Black	9000
P-4	Alamo Beige	21491
P-8	Chatsworth Green	42463
P-I5	Aqua Mist Poly	42462
P-23	Echo Green Poly	426S92
P-26	Valley Green Poly	42813
P-27	Sonata Blue	12360
P-28	Berkeley Blue Poly	12169
P-29	Whirlwind Tan Poly	21846
P-30	Briarcliff Red	71186
P-31	Inca Silver Poly	32087
P-32	Waikiki Gold	81171
P-33	Jasmine Rose	71071
P-34	Fireglow Red Poly	71185
P-72	Frost White	8108
1962 RAMBLER		
P-I	Classic Black	9000
P-I5	Aqua Mist Poly	42462
P-27	Sonata Blue	12360
P-30	Briarcliff Red	71186
P-31	Inca Silver Poly	32087
P-33	Jasmine Rose	71071
P-35	Baron Blue Poly	12499
P-36	Glen Cove Green	42931

P-37	Elmhurst Green Poly	42930
P-38	Algiers Rose Copper Poly	22100
P-39	Villa Red Poly	71240
P-40	Majestic Blue Poly	12498
P-41	Corsican Gold Poly	22099
P-42	Sirocco Beige	22098
P-72	Frost White	8108

1963 RAMBLER

P-1	Classic Black	9000
P-30	Briarcliff Red	71186
P-40	Majestic Blue Poly	12498
P-41	Corsican Gold Poly	22099
P-43	Sceptre Silver Poly	32281
P-44	Bahama Blue	12674
P-45	Cape Cod Blue Poly	12675
P-46	Palisade Green	43098
P-47	Aegean Aqua Poly	43099
P-48	Calais Coral Poly	22257
P-49	Valencia Ivory	81390
P-50	Concord Maroon Poly	50674
P 75	Frost White	8108

1964 RAMBLER

P-1	Classic Black	9000
P-13A	Solar Yellow	81483
P-43	Sceptre Silver Poly	32281
P-43A	Sceptre Silver Poly (Acrylic)	32393
P-51	Rampart Red	71394
P-52A	Sentry Light Blue Poly	12825
P-53	Forum Dark Blue	12826
P-54A	Worlwind Lt. Green Poly	43225
P-55	Westminster Dark Green	43226
P-56	Aurora Light Turquoise	12827
P-S7	Lancelot Medium Turquoise	12828
P-58	Bengal Ivory	81442
P-59	Emperor Light Gold Poly	22362
P-60A	Contessa Rose Poly	71395
P-61	Vintage Maroon Poly	50654
P-72	Frost White	8108

1965 RAMBLER

P-1A	Classic Black	9000,9300
P-3A	Antigua Red	71467
P-4A	Mystic Gold Poly	22508
P-5A	Legion Light Blue	12983
P-6A	Viscount Medium Blue Poly	12984
P-7A	Seaside Light Aqua	12985
P-8A	Marina Medium AquaPoly	12986
P-9A	Atlantis Dark Aqua Poly	12987
P-10A	Montego Light Rose	71468
P-11A	Barcelona Med.Taupe Poly	71469
P-12A	Corral Cordovan Poly	22509
P-13A	Solar Yellow	81483
P-14A	Marlin Silver Poly	32498
P-54A	Worlwide Light Green Poly	43225
P-72A	Frost White	8616

1966 RAMBLER

P-1A	Classic Black	9000,9300
P-3A	Antigua Red	71467
P-15A	Brisbane Light Blue Poly	13131
P-16A	Britannia Dark BluePoly	13132
P-I7A	Crescent Light Green	43488
P-18A	Granada Medium Green Poly	43489
P-19A	Balboa Light Aqua	13133
P-20A	Cortez Medium Aqua Poly	13134
P-21A	Marquessa Light Mauve Poly	50715
P-23A	Samoa Light Gold Poly	22645
P-24A	Caballero Medium Tan Poly	22644
P-25A	Apollo Yellow	81525
P-37A	Sungold Amber Poly	22781
P-72A	Frost White	8616

1967 RAMBLER

P-1A	Black	9300
P-8A	Marina Aqua Poly	12986
P-18A	Granada Green Poly	43489
P-25A	Apollo Yellow	81525
P-31A	Strato Blue Poly	13283
P-32A	Barbados Blue Poly	13584
P-33A	Royal Blue Poly	13285
P-34A	Alameda Aqua	13286
P-36A	Yuma Tan Poly	22765
P-37A	Sungold Amber Poly	22781
P-38A	Stallion Brown Poly	22767
P-39A	Matador Red	71561
P-40A	Flamingo Burgundy Poly	50738
P-41A	Rajah Burgundy Poly	50739
P-42A	Satin Chrome	DX-8555
P-58A	Hialeah Yellow	81606
P-59A	Polo Green Poly	43746
P-72A	Frost White	8616
	Briarcliff Grain Black	DIA-9352*

*(Station Wagon Side Panels)

1968 AMERICAN MOTORS

P-1A	Classic Black	9300
P-39A	Matador Red	71561
P-43A	Saturn Blue Poly	13415
P-44A	Caravelle Blue Poly	13416
P-45A	Blarer Blue Poly	13417

P-47A	Rally Green Poly	43707		P-IO	Bright Blue	13936
P-48A	Tahiti Turquoise Poly	43705		P-88	White	8810
P-49A	Laredo Tan Poly	22877				
P-50A	Calcutta Russet Polv	71599		**1971 AMERICAN MOTORS**		
P-52A	Scarab Gold Poly	22878		A-1	Snow White	2265
P-54A	Turbo Silver Poly	8593		A-2	Canary Yellow	2266
P-59A	Hialeah Yellow	81606		A-4	Skyline Blue	2268
P-72A	Frost White	8616		A-5	Midway Blue Poly	2269
				A-6	Midnight Blue Poly	2270
1969 AMERICAN MOTORS				A-7	Limelight Green Poly	2271
P-39	Matador Red	71561		A-8	Meadow Green Poly	2272
P-62	Ascot Gray	2001		A-8	Raven Green Poly	2273
P-63	Castillian Gray Poly	2002		B-2	Burnished Brown Poly	2274
P-64	Beal St. Blue Poly	2003		B-3	Quick Silver Poly	2275
P-65	Regatta Blue Poly	2004		B-4	Charcoal Gray Poly	2276
P-68	Alamosa Aqua Poly	2005		B-5	Deep Maroon Poly	2277
P-70	Surf Green Poly	2006		B-6	Electric Blue Poly	2244
P-71	Hunter Green Paly	2007		B-7	Brilliant Green Poly	2278
P-72	Frost White	8616		B-8	Mustard Yellow	2279
P-75	Willow Green Poly	2008		B-9	Matador Red	71561
P-76	Pompeii Yellow	2009		C-1	Garden Lime Poly	2123
P-77	Butternut Beige Poly	2010		C-6	Surfside Turquoise	2394
P-78	Cordoba Brown Poly	2011		D-3	Baja Bronze Poly	2398
P-79	Bittersweet Orange Poly	2012		D-9	Wild Rum Poly	2402
P-80	Black Mink Poly	2013		P-1	Classic Black	9000,9300
				P-81	Gray	32353
1970 AMERICAN MOTORS						
B-6	Electric Blue Poly	2244		**1972 AMERICAN MOTORS**		
P-1	Classic Black	9000,9300		A-1	Snow White	2165
P-2	Big Bad Blue	2111		A-2	Canary Yellow	2266
P-3	Big Bad Orange	2114		C-2	Stardust Silver Poly	2390
P-4	Big Bad Green	2115		C-3	Skyway Blue	2391
P-8	Shadow Black	DDL-9381		C-4	Jetset Blue Poly	2392
P-39	Matador Red	71561		C-5	Admiral Blue Poly	2393
P-58	Hialeah Yellow	81606		C-6	Surfside Turquoise	2394
P-72	Frost White	8616		C-8	Grasshopper Green Poly	2395
P-79	Bittersweet Orange Poly	2012		C-9	Hunter Green Poly	2007
P-82	Bayshore Blue Poly	2118		D-1	Jolly Green	2396
P-84	Commodore Blue Poly	2119		D-2	Yuca Tan Poly	2397
P-85	Sea Foam Aqua Poly	2120		D-3	Baja Bronze Poly	2398
P-86	Mosport Green Poly	2121		D-4	Cordoba Brown Poly	2011
P-87	Glen Green Poly	2122		D-5	Butterscotch Gold	2399
P-90	Golden Lime Poly	2123		D-7	Trans-Am Red	2400
P-91	Tijuana Tan Poly	2124		D-8	Sparkling Burgundy	2401
P-94	Moroccan Brown Poly	2125		D-9	Wild Plum Poly	2402
P-95	Sonic Silver Poly	2126		P-1	Classic Black	9000,9300
P-81	Gray	32353		P-81	Gray	32353
1970 S/C RAMBLER COLORS				**1973 AMERICAN MOTORS**		
P-7	Flat Black	9378		A-1	Snow White	2265
P-9	Bright Red	71816		C-8	Grasshopper Green Poly	2395

C-8-A	Grasshopper Green Poly	2587
D-4	Cordoba Brown Poly	2011
D-7	Trans-Am Red	2400
E-1	Diamond Blue Poly	2484
E-1-A	Diamond BluePoly	2588
E-2	Olympic Blue Poly	2485
E-3	Fairway Green Poly	2486
E-4	Tallyho Green Poly	2487
E-5	Pewter Silver Poly	2488
E-5-A	PewterSilver Poly	2589
E-6	Fawn Beige	2489
E-7	Copper Tan Poly	2490
E-9	Mellow Yellow	2492
F-1	Blarney Green	2493
F-2	Maxi Blue	2494
F-3	Fresh Plum Poly	2495
F-4	Daisy Yellow	2496
F-5	Vinyard Burgundy Poly	2506
P-1	Classic Black	9000,9300
P-81	Gray	32353

1974 AMERICAN MOTORS

A-7	Snow White	2597
D-7	Trans-Am Red	2400
E-1-A	Diamond Blue Poly	2588
E-6	Fawn Beige	2489
E-9	Mellow Yellow	2492
F-2	Maxi Blue	2494
F-4	Daisy Yellow	2496
F-5	Vinyard Burgundy Poly	2506
F-6	Medium Blue Poly	2598
F-7	Dark Blue Poly	2599
F-8	Golden Tan Poly	2600
F-9	Copper Poly	2601
G-1	Silver Green Poly	2602
G-2	Medium Green Poly	2603
G-3	Dark Green Poly	2604
G-4	Plum Poly	2605
G-5	Pewter Mist Poly	2606
G-6	Sienna Orange	2607
J-7	Ivory Green	2732
J-8	Caramel Tan	2733
P-1	Classic Black	9000,9300
P-81	Gray	32353

1975 AMERICAN MOTORS

D-7	Trans-Am Red	2400
E-6	Fawn Beige	2489
E-9	Mellow Yellow	2492
F-9	Copper Poly	2601
G-3	Dark Green Poly	2604

G-6	Sienna Orange	2607
G-7	Alpine White	2702
G-8	Pastel Blue	2703
G-9	Medium Blue Poly	2704
H-1	Deep Blue Poly	2705
H-2	Reef Green Poly	2706
H-4	Dark Cocoa Poly	2707
H4	Dark Cocoa Poly	2796
H-5	Green Apple	2708
H-6	Golden Jade Poly	2709
H-7	Aztec Copper Poly	2710
H-8	Autumn Red Poly	2711
H-9	Silver Dawn Poly	2712
J-2	Brandywine Poly	2714
J-7	Ivory Green	2732
J-8	Caramel Tan	2733
P-1	Classic Black	9000,9300
P-81	Gray	32353
6A	Marine Aqua Poly	2817
6B	Seaspray Green	2818
6C	Evergreen Poly	2819
6D	Sand Tan	2820
6E	Burnished Bronze Poly	2821
6J	Silver Frost Poly	2822
6K	Limefire Poly	2823
6P	Firecracker Red	2824
6R	Brilliant Blue	2825
6T	Nautical Blue Poly	2826
6V	Sunshine Yellow	2827

1976 AMERICAN MOTORS

G-6	Sienna Orange	2607
G-7	Alpine White	2702
G-9	Medium Blue Poly	2704
F-9	Copper Poly	2601
H-4	Dark Cocoa Poly	2796
H-6	Golden Jade Poly	2709
H-7	Aztec Copper Poly	2710
H-8	Autumn Red Poly	2711
J-2	Brandywine Poly	2714
6-A	Marine Aqua Poly	2817
6-B	Seaspray Green	2818
B-C	Evergreen Poly	2819
6-D	SandTan	2820
6-E	Burnished Bronze Poly	2821
6-J	Silver Frost Poly	2822
6-K	Limefire Poly	2823
6-P	Firecracker Red	2824
6-R	Brilliant Blue	2825
6-T	Nautical Blue Poly	2826
6-V	Sunshine Yellow	2827
P-I	Classic Black	9000,9300

9

1977 AMERICAN MOTORS

G7	Alpine White	2702
J1	Pewter Gray Poly	33160
J2	Brandywine Poly	2714
6D	Sand Tan	2820
6J	Silver Frost Poly	2822
6P	Firecracker Red	2824
6R	Brilliant Blue	2825
6V	Sunshine Yellow	2827
7A	Misty Jade Poly	2894
7B	Mocha Brown Poly	2895
7C	Autumn Red Poly	2992
7D	Powder Blue	2896
7E	Oak Leaf Brown	24443
7K	Midnight Blue Poly	2897
7L	Loden Green Pnly	2898
7M	Golden Ginger Poly	2899
7P	Lime Green	3000
7W	Captain Blue Poly	2901
7Y	Tawny Orange	2902
7Z	Sun Orange	2903
PI	Classic Black	9000,9300

1978 AMERICAN MOTORS

G-7	Alpine White	2702
J-1	PewterGray Poly	33160
6-D	Sand Tan	2820
6-P	Firecracker Red	2824
6-R	Brilliant Blue	2825
6-V	Sunshine Yellow	2827
7-B	MochaSrown Poly	2895
7-C	Autumn Red Poly	2992
7-D	Powder Blue	2896
7-E	Oak Leaf Brown	24443
7-K	Midnight Blue Poly	2897
7-L	Loden Green Poly	2898
7-M	Golden Ginger Poly	2899
7-W	Captain Blue Poly	2901
7-Z	Sun Orange	2903
8-A	Khaki	3028
8-B	British Bronze Poly	3029
8-C	Quicksilver Poly	3030
8-D	Claret Poly	3022
P-1	Classic Black	9000,9300

1979 AMERICAN MOTORS

8P	Firecracker Red	2824
8A	Khaki	3028
8B	British Bronze Poly	3029
8C	Quick Silver Poly	3030
9A	Alpaca Brvwn Poly	3106
9B	Olympic White	3107
9C	Russet Poly	3108
9E	Wedgwood Blue	3109
9H	Cumberland Green Poly	3110
9K	Sable Brown Poly	3112
9L	Saxon Yellow	3113
9M	Starboard Blue Poly	3114
9N	Morocco Buff	3115
9P	Bordeaux Poly	3111
9Z	Misty Beige Poly	24635*
P1	Black	9300

*Pacer Only

1980 AMERICAN MOTORS

8C	Quick Silver Poly	3030
9B	Olympic White	3107
9C	Russet Poly	3108
9L	Saxon Yellow	3113
9P	Bordeaux Poly	3111
	(Pacer)	
9Z	Misty Beige Poly	24635
OB	Smoke Gray Poly	3239
OC	Cameo Blue	3240
OD	Medium Blue Poly	3241
OE	Dark Green Poly	32420H
	(Concord)	
	Navy Blue	3243
OJ	Teal Blue	3244
OK	Cameo Tan	3245
OL	Medium Brown Poly	3246
OM	Dark Brown Poly	3247
OP	Cardinal Red	3248
OR	Caramel	32490T
	(Spirit, Pacer)	
	Black Poly (Medium Gloss)	9429
	(Spirit, Concord)	
PI	Black	9300

1981 AMERICAN MOTORS

8C	Quick Silver Poly	3030
9B	Olympic White	3107
OD	Medium Blue Poly	3241
OK	Cameo Tan	3245
OL	Medium Brown Poly	3246
OM	Dark Brown Poly	3246
1B	Moonlight Blue	3367
1C	Sherwood Green Poly	3368
1D	Autumn Gold	3369
1E	Copper Brown Poly	3370
1J	Vintage Red Poly	3372
1K	Deep Maroon Poly	3373

1L	Steel Gray Poly	3374
1M	Oriental Red	3375
P1	Black	9300

1982 AMERICAN MOTORS

9B	Olympic White	3107OM(S,C,E,X)
	Dark Brown Poly	3247(C,E,X)
OT	Black (Accent)	9429
1C	Sherwood Green Poly	3368(C)
1E	Copper Brown Poly	3370(S,C,E,X)
1J	Vintage Red Poly	3372(E)
1K	Deep Maroon Poly	3373(S,C,E,X)
1M	Oriental Red	3375(S,E,X)
2A	Mist Silver Poly	3466(S,C,E,X)
2B	Sun Yellow	3467(S,X)
2C	Slate Blue Poly	3468(S,C,E,X)
2D	Deep Night Blue	3469(S,C,E,X)
2E	Sea Blue Poly	3470((S,C)
2H	Topaz Gold Poly	3471(S,C,E,X)
2J	Jamaican Beige	3472(S,C,E,X)
P1	Black	9300(S,C,E,X)

S-Spirit, C-Concord, E-Eagle, X-SX4

1983 AMERICAN MOTORS

OM	Dark Brown Poly	3247(E,S)
1E	Copper Brown Poly	3370(E,S,C)
1J	Vintage Red Poly	3372(E,S,C)
1K	Deep Maroon Poly	3373(E,S,C)
2A	Mist Silver Poly	3466(E,S,C)
2C	Slate Blue Poly	3468(E,S,C)
2D	Deep Night Blue	3469(E,S,C)
2H	Topaz Gold Poly	3471(E,S,C)
2J	Jamaican Beige	3472(E,S,C)
3B	Sebring Red	3685(E,S,C)
9B	White	3107(E,S,C)
P1	Black	9300(E,S,C)
OT	Black Accent	9429(E,S)

C-Concord, S-Spirit, E-Eagle

1984 AMERICAN MOTORS

2A	Mist Silver Poly	3466
2C	Slate Blue Poly	3468
2D	Deep Night Blue	3469
2H	Topaz Gold Poly	3471
3B	Sebring Red	3592
3P	Garnet Poly	3599
4A	Autumn Brown Poly	3622
4B	Mocha Brown Poly	3623
4E	Chestnut Brown Poly	3626
9B	White	3107
P1	Black	9300
OT	Black Accent	9429

1985 AMERICAN MOTORS

2A	Mist Silver Poly	3466
3A	Almond Beige	3591
3B	Sebring Red	3592
3D	Sterling Poly	3594
4A	Autumn Brown PolV	3622
4B	Mocha Brown Poly	3623
5H	Medium Blue Poly	3698
5J	Dark Blue Poly	3699
9B	Olympic White	3700
P1	Classic Black	9300

1986 AMERICAN MOTORS

2A	Mist Silver Poly	3466
3C	Garnet Poly	3599
4A	Autumn Brown Poly	3622
4B	Mocha Dk. Brown Poly	3623
5H	Medium Blue Poly	3698
5J	Dark Blue Poly	3699
AB	Beige	3826
9B	Olympic White	3107
P1	Classic Black	9300

1987 AMERICAN MOTORS

2A	Mist Silver Poly	3466
3B	Sebring Red	3592
3P	Garnet Poly	3599
4A	Autumn Brown Poly	3622
4B	Mocha Dk. Brown Poly	3623
5H	Med. Blue Poly	3698
5J	Dk. Blue Poly	3699
P1	Classic Black	9000/9300
9B	Olympic White	3107

1988 AMERICAN MOTORS

3D	Sterling Poly	3594
4B	Mocha Brown Poly	3623
5J	Dark Blue Poly	3699
BH	Silver Poly	4115
BK	Vivid Red Poly	4117
CA	Pearl White	4026
CA	Pearl White	4027
CB	Buff Yellow	4028
CC	Dover Gray Poly	4029
P1	Classic Black	9700

Chapter 2   BUICK 1941-1978

PAINT CODE	COLOR	PPG CODE
560	Black	9200
561	Rainier Blue	10070
562	Verde Green Dk.	40072
563	Royal Maroon Poly	50014
564	Monterey Blue Poly	10071
565	Lancaster Gray Poly	30111
566	Chenanga Gray Poly	30058
567	Touquet Beige	30110
568	Sienna Rust	70033
569	Ridge Green Poly	40046
570	Ludington Green	40071
577	English Green	40073
578	Titian Maroon	50015
580	Mermaid Green	40088
581	Cedar Green	40089
584	Yaikima Gray Poly	30079
	Silver French Gray Poly	30051
	Pearl Gray Poly	30071

TWO-TONES

572	U 30051	575	U 40071	582	U 10095
	L 30111		L 40073		L 30079
573	U 30058	576	U 30071	585	U 30079
	L 40046		L 10071		L 10095
574	U 30071	579	U 40088		
	L 9200		L 40089		

1942 BUICK

4201	Carlsbad Black	9200
4202	Ranier Blue	10070
4203	Nightshade Blue Poly	10095
4204	Honolulu Blue Poly	10109
4205	Verde Green	40104
4206	Salt Lake Green	40105
4207	Ludington Green	40071
4208	English Green	40106
4210	Yakima Gray Poly	30079
4211	Lancaster Gray Poly	30111
4212	Folkstone Gray	30092
4213	Chenanga Gray Poly	30058
4214	Royal Maroon Poly	50014
4215	Seguoia Beige	30093
4216	Permanent Red	70030
4217	Dusty Gray	30066
4218	Rivermist Gray	30069
	Silver Gray French Poly	30051
	Pearl Gray Poly	30

TWO-TONES

4219	U 30051	4223	U 30094	4227	U 30069
	L 30111		L 10109		L 30066
4220	U 30058	4224	U 10095	4228	U 30066
	L 40107		L 30079		L 30069
4221	U 30094	4225	U 30079		
	L 9200		L 10095		
4222	U 40071	4226	U 30093		
	L 40106		L 30079		

1946 BUICK

4601	Black	9200
4602	Nightshade Blue Poly	10095
4603	Cantebury Blue Poly	10246
4604	Verde Green	40104
4605	Sheerwood Green Poly	40275
4606	Brunswick green Poly	40276
4608	Dusty Gray	30066
4609	Rivermist Gray	30069
4613	Lehigh Gray Poly	30323
4614	Sequoia Cream	80118

TWO-TONES

4610	U 30069	4611	U 40276
	L 30066		L 40275

1947 BUICK

4701	Black	9200
4704	Verde Green	40104
4705	Sherwood Green Poly	40275
4706	Brunswick Green Poly	40276
4707	Royal Maroon Poly	50014
4714	Sequoia Cream	80118
4715	Seine Blue Poly	10305
4717	Calvert Blue Poly	10413
4718	Regency Blue Poly	10417
4719	Catalina Gray	30458
	Lehigh Gray Poly	30323

TWO-TONES

4711	U 40276	4720	U 30323	4721	U 10417
	L 40275		L 30458		L 30458

1948 BUICK

4801	Carlsbad Black	9200
4804	Verde Green	40104
4807	Royal Maroon Poly	50014
4814	Sequoia Cream	80118

4817	Calvert Blue Poly	10413
48I8	Regency BIlue Poly	10417
4823	Nickel Gray Poly	30504
4824	Honolulu Blue Poly	10109
4825	Aztec Green Poly	40567
4826	Allendale Green Poly	40568
4828	Cumulus Gray	30503
4830	Royal Maroon Poly	50125

TWO-TONES

4822	U 40568	4829	U 30504
	L 40567		L 30503

1949 BUICK

4902	Bahama Blue Poly	10556
4903	Sunmist Gray Poly	30623
4904	Verde Green	40104
4905	Elan Blue Poly	10557
4907	Sequoia Cream	80118
4908	Cumulus Gray	30503
4909	Regency Blue Poly	10417
4910	Allendale Green	40568
4910	Gala Green Poly	40702
4912	Royal Maroon #2 Poly	50125
4913	Cirrus Green Poly	40711
4814	Old Ivory	80433
4916	Mariner Blue Poly	10669
4916	Mariner Blue Poly	10610

I95O BUICK

5001	Black	9200
5002	Cumberland Gray Poly	30739
5003	Verde Green	40104
5004	Imperial Blue Poly	10674
5005	Sunmist Gray Poly	30623
5006	Allandale Green	40568
5007	Royal Maroon #2 Poly	50125
5008	Cirrus Green Poly	40711
5009	Old Ivory	80433
5010	Olympic Blue Poly	10675
5014	Geneva Green Poly	40924
5015	Cumberland Gray	30759
5017	Niagara Green	40941
5018	Calvin Gray	30775
5019	Barton Gray	30774
5020	Meredith Creen	40942
5024	Cloudmist Gray Poly	30868
5025	Kasmir Green Poly	40989

TWO-TONES

5011	U 40104	5021	U 40104

	L 40711		L 40941
5012	U 10674	5022	U 9200
	L 10675		L 30775
5013	U 30623	5023	U 10674
	L 30739		L 30774
5016	U 30623		
	L 30759		

1951 BUICK

5101	Carlsbad Black	9200
5102	Verde Green	40104
5103	Imperial Blue Poly	10674
5104	Geneva Green Poly	40924
5105	Barton Gray	30774
5106	Olympic Blue Poly	10675
5107	Victoria Maroon Poly	50229
5108	Sharon Green Poly	41079
5109	Cloudmist Gray Poly	30868
5110	Old Ivory	80433
5119	Calumet Green Poly	40560
5121	Venetian Blue Polv	10587
5123	Galena Blue Poly	10779
5126	Glenn Green	40561
	Sky Gray	30881

TWO-TONES

5111	U 30881	5120	U 30881
	L 40104		L 40560
5112	U 30881	5122	U 30881
	L 10675		L 10567
5113	U 30881	5124	U 30881
	L 502Z9		L 10779
5114	U 30881	5125	U 10674
	L 41079		L 10779
5115	U 30881	5128	U 40560
	L 30868		L 40561
5116	U 10674		
	L 30774		

1952 BUICK

5201	Black	9200
5202	Verde Green Poly	41162
5203	Imperial Blue Poly	10674
5204	Barton Gray	30774
5205	Victoria Maroon Poly	50229
5206	Seamist Gray Poly	30946
5207	Sky Gray	30881
5208	Terrace Green Sympho	41163
5209	Venetian Blue Poly	10567
5210	Surf Blue	10843
5211	Glenn Green	40561

5212	Sequoia Cream	80118
5213	Apache Red	70389
5228	Nassau Blue Poly	10875
5229	Golden Sand Poly	20845
5230	Coronet Copper Poly	20846
5231	Glacier Green Poly	41204
5232	Peacock Green Poly	41205
5233	Aztec Gold Poly	20847
5240	Golden Sand Poly	20878
5242	Teal Blue Poly	10924
5243	Beach White	20820

TWO-TONES

5214	U 30881 Top		5225	U Convertible
	L 41162			L 70389
5215	U 30881		5234	U 20820
	L 50229			L 10875
5216	U 30881		5235	U 20820
	L 30946			L 20845
5217	U 10674		5236	U 20820
	L 30774			L 20846
5219	U 30881		5238	U 20820
	L 41163			L 41205
5220	U 30881		5239	U 41162
	L 10567			L 20847
5221	U 30881		5241	U 20820
	L 10843			L 20878
5222	U 10674		5244	U 20820
	L 10843			L 10924
5224	U Convertable Top		5245	U 30946
	L Sequoia Cream			L 20820

1953 BUICK

5351	Carlsbad Black	9200
5352	Verde Green Poly	41162
5353	Imperial Blue Poly	10674
5354	Jordan Gray	31078
5355	Victoria Maroon Poly	50229
5356	Seamist Green Poly	30946
5357	Shell Gray	31079
5358	Terrace Green Sympho	41163
5359	Tyler Blue Poly	10959
5360	Ridge Green	41134
5362	Matador Red	70422
5373	Teal Blue Poly	10924
5374	Majestic White	31080
5378	Mandarin Red Poly	50339
5379	Balsam Green Poly	41335
5383	Pinehurst Green Poly	41336
5386	Glacier Blue	10929

TWO-TONES

5363	U 31079		5375	U 31080
	L 41162			L 10924
5364	U 31079		5376	U 30946
	L 50229			L 31080
5365	U 31079		5380	U 31080
	L 30946			L 41335
5366	U 10674		5381	U 31080
	L 31078			L 50339
5367	U 30946		5382	U 31080
	L 31079			L 10960
5368	U 41334		5384	U 31080
	L 41163			L 41336
5369	U 31079		5385	U 31080
	L 10959			L 70422
5370	U 41163		5387	U 10674
	L 41334			L 10929
5371	U 41162		5388	U 31078
	L 80614			L 10929
5372	U 9200			
	L 70422			

1954 BUICK

5401	Carlsbad Black	9200
5402	Artic White	8093
5403	Casino Beige	21092
5405	Gull Gray Poly	31195
5408	Cavalier Blue Poly	11181
5409	Ranier Blue Poly	11182
5410	Marlin Blue	11183
5411	Malibu Blue	11180
5412	Baffin Green Poly	41603
5413	Willow Green	41602
5414	Ocean Mist Green	41606
5415	Aztec Green Poly	41605
5416	Lido Green	41604
5417	Titian Red Poly	70500
5418	Matador Red	70422
5420	Condor Yellow	80691

1955 BUICK

AAA	Carlsbad Black	9200
BBB	Dover White	8030
CCC	Cameo Beige	21167
EEE	Temple Gray	31270
FFF	Colonial Blue Poly	11387
GGG	Victoria Blue Poly	11343
HHH	Cascade Blue	11342
KKK	Stafford Blue Poly	11344
LLL	Belfast Green Poly	41753
NNN	Galway Green	41754

SSS	Cherokee Red	70584
TTT	Gulf Turquoise	41624
WWW	Spruce Green Poly	41811
XXX	Nile Green Poly	41812
YYY	Mist Green	41731

1956 BUICK

56A	Carlsbad Black	9200
56B	Castle Gray Poly	31375
56C	Dover White	8030
56D	Electric Blue Poly	11503
56E	Bedford Blue	11502
56F	Cadet Blue Poly	11386
56G	Cambridge Poly	11501
56H	Laurel Green Poly	41942
56J	Foam Green	41940
56K	Glacier Green	41941
56L	Claret Red Poly	50442
56M	Seminole Red	70680
56N	Tahiti Coral	70679
56P	Cameo Beige	21167
56R	Harvest Yellow	80829
56T	Bittersweet	60240
56U	Apricot	60239

1957 BUICK

C	Dover White	8030
D	Starlight Blue Poly	11622
E	Biscay Blue Poly	11623
F	Mariner Blue Poly	11702
H	Kearney Green Poly	42076
J	Belmont Green	42077
K	Mint Green	42078
L	Jade Green	42075
N	Garnet Red Poly	60253
R	Antique Ivory	80904
S	Artic Blue	11701
T	Dawn Gray	31453
U	Gulf Green Poly	42174
Y	Dusk Rose	50488

1958 BUICK

58A	Carlsbad Black	9200
58B	Sylvan Gray Poly	31579
58C	Glacier White	8166
58D	Hunter Green Poly	42175
58E	Spray Green	42220
58H	Dark Turquoise Poly	11774
58K	Antique Ivory	80904
58L	Casino Cream	80973
58M	Seminole Red	70851

58N	Garnet Red Poly	60253
58P	Reef Coral	70748
58R	Desert Beige	21541
58S	Cobalt Blue	11964
58T	Desert Sage	50527
58U	Mohave Yellow	80902
58W	Canyon Cedar	70936
58-I	Green Mist	42250
58-2	Blue Mist	11773
58-3	Laurel Mist	50507
58-4	Silver Mist	31458
58-5	Polar Mist	8155
58-6	Gold Mist Poly	21694

1959 BUICK

59A	Sable Black	9200
59B	Silver Birch Poly	31827
59C	Artic White	8160
59D	Sierra Brown Poly	42479
59E	Glacier Green Poly	42480
59H	Shalimar Blue Poly	12002
59J	Wedgewood Blue	12003
59K	Turquoise Poly	12001
59L	Tampico Red	70961
59M	Tawny Rose Poly	70959
59N	Lido Lavender Poly	50536
59P	Pearl Fawn Poly	21722
59R	Copper Glow Poly	21723

1960 BUICK

AA	Sable Black	9300
BB	Gull Gray Poly	31905
CC	Arctic White	8259
DD	Silver Mist Poly	31928
HH	Chalet Blue Poly	12234
KK	Lucerne Green Poly	42693
LL	Titian Red Poly	50568
MM	Casino Cream	81202
NN	Cordovan Poly	21874
PP	Pearl Fawn Poly	21722
RR	Tahiti Beige	21873
TT	Turquoise Poly	12228
VV	Tampico Red	70961
WW	Midnight Blue Poly	12224
XX	Verde Green Poly	42696

Two-Tone #1- Lower body is basic color, roof color carries accent color.

Two-Tone #2-Roof and upper wing areas carry accent color,balance of car is basic color

Two-Tone #3-Roof upper wing area and below the body

side molding carry the two-tone or accent color. Balance of car is basic color.

1961 BUICK

A	Sable Black	9300
C	Arctic White	8259
D	Newport Silver Poly	31928
E	Venice Blue Poly	12397
F	Laguna Blue Poly	12398
H	Bimini Blue	12399
J	Dublin Green Poly	42837
K	Kerry Green	42838
L	Rio Red Poly	50568
M	Sun Valley Cream	81271
N	Cordovan Poly	21874
P	Turquoise Poly	12396
R	Phoenix Beige	21733
T	Desert Fawn Poly	22005
V	Tampico Red	70961

TWO-TONES: First letter is basic body color, second letter is accent color. Example- AC
If a number is used after tbe letters the following applies:
SPECIAL: 1-Top only is two tone.
Example: AC-1
Black body with white top.
LASABRE: 1-Top only is two-tone. 2-Top, hood, upper front fenders, center section of body and deck lid is the accent color.
INVICTA: 1-Roof only.
(Upper hody side peak moulding,
lower body slde peak moulding.)
ELECTRA: 1-Roof and rear deck panel below trunk lid.

1962 BUICK

A	Regal Black	9300
C	Arctic White	8259
D	Silver Cloud Poly	32173
E	Cadet Blue Poly	12552
F	Marlin Blue Poly	12546
H	Glacier Blue	12549
J	Willow Mist Poly	42975
M	Cameo Cream	81271
N	Burgundy Poly	50615
P	Teal Mist Poly	12525
Q	Aquamarine	I2550
R	Desert Sand	22137
T	Fawn Mist Poly	22121
V	Cardinal Red	70961
X	Camelot Rose Poly	71269

TWO-TONES: The first letter is the lower body color and the second letter the roof color.

1963 BUICK

A	Regal Black	9800
C	Arctic White	8259
D	Silver Cloud Poly	32173
E	Spruce Green Poly	43125
F	Marlin Blue Poly	12546
H	Glacier Blue	12713
J	Willow Mist Poly	42975
N	Burgundy Poly	50640
P	Teal Mist Poly	12525
Q	Twilight Aqua Poly	43114
R	Desert Sand	22137
S	Bronze Mist Poly	22269
T	Fawn Mist Poly	22268
V	Granada Red	71336
W	Diplomat Blue Poly	12696
X	Rose Mist Poly	71337

TWO-TONES:
The first letter is the lower body color and the second letter the roof color.

1964 BUICK

A	Regal Black	9300
C	Arctic White	8259
D	Silver Cloud Poly	32173
F	Marlin Blue Poly	12546
H	Wedgewood Blue	12847
J	Surf Green Poly	43264
L	Claret Mist Poly	50633
L	Claret Mist Poly #2	50684
N	Coral Mist Poly	71415
P	Teal Mist Poly	12525
R	Desert Beige	22391
S	Bronze Mist Poly	22269
T	Tawny Mist Poly	22392
V	Granada Red	71336
W	Diplomat Blue Poly	12696
	Accent Strip-Argent Silver	8568

TWO-TONES: The first letter is the lower body color and the second letter the upper color.

1965 BUICK

A	Regal Black	9300
C	Arctic White	8259
D	Astro Blue Poly	13042
E	Midnight Bluc Poly	13002

H	Seafoam Green Poly	43391
J	Verde Green Poly	43390
K	Turquoise Mist Poly	43364
L	Midnight Aqua Poly	13003
N	Burgundy Mist Poly	50700
R	Flame Red	71472
S	Sahara Mist Poly	22553
T	Champagne Mist Poly	22564
V	Shell Beige	22270
Y	Bamboo Cream	81500
Z	Silver Cloud Poly	32173

TWO-TONES: The first letter is the lower body color and the second letter the upper color.

1966 BUICK

A	Regal Black	9300
B	Riviera Gunmetal Poly	32448
C	Arctic White	8259
D	Astro Blue Poly	13042
E	Midnight Blue Poly	13002
F	Blue Mist Poly	13148
G	Riviera Gold Poly	22661
H	Seafoam Green Poly	43391
J	Verde Green Poly	43390
K	Turquoise Mist Poly	43364
L	Shadow Turquoise Poly	43496
M	Riviera Red Poly	71525
N	Burgundy Mist Poly	50700
R	Flame Red	71472
S	Riviera Champagne Poly	22662
T	Saddle Mist Poly	22660
U	Riviera Plum	50722
V	Shell Beige	22270
W	Silver Mist Poly	32525
X	Riviera White	8631
Y	Cream	81528
Z	Riviera Silver Green Poly	43525
	White- Semi-gloss	8659*
	Black- Semi-gloss	9339*

*Riviera rear compartment exhaust grille

TWO-TONES: The first letter is the lower body color and the second letter the upper color.

1967 BUICK

A	Regal Black	9300
B	Riviera Turquoise Poly	43664*
C	Artic White	8259
D	Sapphire Blue Poly	13349
E	Midnight Blue Poly	13346
F	Blue Mist Poly	13364
G	Gold Mist Poly	22818
H	Green Mist Poly	43651
J	Verde Green Poly	43653
K	Aquamarine Poly	43661
L	Shadow Turquoise Poly	43659
N	Burgundy Mist Poly	50700
P	Platinum Mist Poly	32603
R	Apple Red	71583
S	Champagne Mist Poly	22813
T	Ivory	81578
U	Riviera Plum	50722*
V	Riviera Charcoal Poly	32604*
W	Riviera Fawn Poly	22821*
X	Riviera Red Poly	71585*
Z	Riviera Gold Poly	43665*
L	Smoker's Lung Gray	12345

* Riviera Colors

	White- Semi-gloss	8659*
	Black- Semi-gloss	9339*
	Dk. Blue- Semi gloss	13456*
	Dk. Green- Semi-gloss	43728*

*Riviera rear compartment exhaust grille

TWO-TONES: The first letter is the lower body color and the second letter the upper color.

1968 BUICK

A	Regal Black	9300
B	Midnight Teal Poly	13515*
C	Arctic White	8259
D	Blue Mist Poly	13512
E	Deep Blue Poly	13513
F	Teal Blue Mist Poly	13514
G	Ivory Gold Mist Poly	22942
K	Aqua Mist Poly	13517
L	Medium Teal Blue Mist Poly	13516
M	Burnished Saddle Poly	22967
N	Maroon Poly	50775
P	Tarpon Green Mist Poly	43774
R	Scarlet Red	71634
S	Olive Gold Poly	43794*
T	Desert Beige	81617
V	Charcoal Poly	43773
W	Silver Beige Mist Poly	22962*
X	Buckskin	22983*
Y	Cameo Cream	81500
Z	Inca Silver Mist	8596*

*Riviera colors

TWO-TONES: The first letter is the lower body color, the second letter the upper color.

1969 BUICK

10	Regal Black	9300
40	Cameo Cream	81500
50	Polar White	2058*
51	Twilight Blue Poly	2075
52	Signal Red	2076*
53	Crystal Blue Poly	2077
55	Turquoise Mist Poly	2078
57	Verde Green Poly	2079
59	Lime Green Poly	20B0
61	Burnished Brown Poly	2081
63	Champagne Mist Poly	22813
65	Trumpet Gold Poly	2082
67	Burgundy Mist Poly	50700
69	Silver Mist Poly	2059
75	Embassy Gold Poly	2085*
77	Antique Gold Poly	2086**
80	Azure Blue	2096**
81	Sunset Silver Poly	2087**
82	Olive Beige	2088**
83	Deep Gray Mist Poly	32526**
85	Copper Mist Poly	22683**
Q	Fire Glow Orange Poly	
	60586(Spring Color)	

*Striping Colors
**Riviera Colors

TWO-TONES: The first two digits are the lower body color, the second two digits the upper color.

1970 BUICK

10	Glacier White	8631
14	Silver Mist Poly	2059*
16	Tealmist Gray Poly	2161**
19	Regal Black	9300
20	Azure Blue	2162**
25	Gulfstream Blue Poly	2165*
26	Stratomist Blue Poly	2213**
28	Diplomat Blue Poly	2166*
34	Aqua Mist Poly	2168*
45	Seamist Green Poly	2171*
46	Emerald Mist Poly	2172**
48	Sherwood Green Poly	2173*
50	Bamboo Cream	2175
55	Cornet Gold Poly	2178*
58	Harvest Gold Poly	2179*
61	Sandpiper Beige	2181**
63	DesertGold Poly	2183*

68	Burnished Saddle Poly	2233*
74	Titian Red Poly	71642**
75	Fire Red	2189*
76	Sunset Sage Poly	2190**
78	Burgundy Mist Poly	50700*
—	Saturn Yellow	
	81838(Buick GSX)	

*Exterior Wheel Colors
**Riviera Colors

TWO-TONES: The first two digits are the lower body color, the second two digits the upper color.

1971 BUICK

11	Arctic White	2058
13	Platinum Mist Poly	2327
16	Tealmist Gray Poly	2161
19	Regal Black	9300
24	Cascade Blue Poly	2328
26	Stratomist Blue Poly	2213
29	Nocturne Blue Poly	2330
39	Twilight Turquoise Poly	2331
41	Silver Fern Poly	2332
42	Willomist Green Poly	2333
43	Lime Mist Poly	2334
50	Bamboo Cream	2175*
53	Cortez Gold Poly	2339
55	Cornet Gold Poly	2178
61	Sandpiper Beige	2181
62	Bittersweet Mist Poly	2340
65	Copper Mist Poly	2343*
67	Burnished Cinnamon Poly	23215
68	Deep Chestnut Poly	2344*
70	Pearl Beige Poly	2346*
73	Sunset Mist Poly	2347*
74	Vintage Red Poly	2348*
75	Fire Red	2189
78	Rosewood Poly	2350

*Riviera Colors

TWO-TONES: The first two digits are the lower body color, the second two digits the upper color.

1972 BUICK

11	Artic White	2058
19	Regal Black	9300
14	Silver Mist Poly	2429
18	Charcoal Mist Poly	2430
21	Crystal Blue Poly	2431
24	Cascade Blue Poly	2328
26	Stratomist Blue Poly	2213

28	Royal Blue Poly	2166
36	Heritage Green Poly	2433
41	Seamist Green Poly	2435
45	Emerald Mist Poly	2172
48	Hunter Green Poly	2439
50	Covert Tan	2441
53	Cortez Gold Poly	2339
54	Champagne Gold Poly	2442
56	Sunburst Yellow	2444
57	Antique Gold Poly	2445
62	Sierra Tan	2447
63	Burnished Copper Poly	2448
65	Flame Orange Poly	2450
67	Deep Chestnut Poly	2344
69	Nutmeg Poly	2452
73	Vantage Red Poly	2348
75	Fire Red	2189
77	Burnished Bronze Poly	2454

TWO-TONES: The first two digits are the lower body color, the second two digits the upper color.

1973 BUICK

11	Arctic White	2058
19	Regal Black	9300
24	Medium Blue Poly	2523
26	Mediterranean Blue Poly	2524
29	Midnight Blue Poly	2526
42	Jade Green Poly	2528
44	Willow Green Poly	2529
46	Green-Gold Poly	2530
48	Midnight Green	2531
51	Apollo Yellow	2533
54	Autumn Gold	2536*
56	Colonial Yellow	2537
60	Harvest Gold Poly	2538
64	Silver Cloud Poly	2541
65	Burnt Coral Poly	2557*
66	Taupe Poly	2542
67	Midnight Gray Poly	2558*
68	Brown Poly	2543
74	Burgundy Poly	2545
75	Apollo Red	2546
81	Bamboo Cream	2549
97	Apollo Orange	2555
99	Gall Bladder Green	2667

*Riviera

TWO-TONES: The first two digits are the lower body color, the second two digits the upper color.

1974 BUICK

11	Artic White	2058
19	Regal Black	9300
24	Medium Blue Poly	2523
26	Mediterranean Blue Poly	2524
29	Midnight Blue Poly	2526
36	Crystal Lake Blue Poly	2640
40	Mint Green	2641
44	Ranch Green	2642
46	Leaf Green Poly	2643
49	Forest Green Poly	2645
50	Sand Beige	2646
51	Canary Yellow	2677
53	Ginger Poly	2649
54	Gold Mist Poly	2442
55	Nugget Gold	2650
59	Nutmeg Poly	2367
64	Silver Cloud Poly	2541
66	Cinnamon Poly	2653
69	Dark Brown Poly	2656
72	Ruby Red	2544
74	Burgundy Poly	2658
75	Apple Red	2546
79	Plum Poly	2659

TWO-TONES: The first two digits are the lower body color, the second two digits the upper color.

1975 BUICK

11	Arctic White	2058
13	Silver Mist	2518
15	Dove Gray	2742
16	Pewter	2743
19	Regal Black	9300
18	Antique Silver	2771*
21	Horizon Blue	2431
24	Glacier Blue	2745
26	Majestic Blue	2746
28	Blue Haze	2772*
29	Indigo	2748
44	Ranch Green	2642
49	Verde Mist	2752
50	Sand Beige	2646
51	Canary Yellow	2677
55	Sandstone	2756
57	Honey Gold	2649*
58	Almond Mist	2757
59	Walnut Mist	2758
63	Golden Tan	2759
64	Bittersweet	2760
72	Ruby Red	2544

74	Burgundy	2658
75	Apple Red	2546
79	Rhone Red	2659
80	Pumpkin	2548

*Riviera

TWO-TONES: The first two digits are the lower body color, the second two digits the upper color.

1976 BUICK

11	Liberty White	2058
13	Pewter Gray Poly	2518
16	Medium Gray Poly	2862
19	Judical Black	9300
28	Potomac Blue Poly	2772
35	Continental Blue	2863
37	Independence Red Poly	2864
40	Concord Green Poly	2866
49	Dark Green Poly	2752
50	Mt.Vernon Cream	2867
51	Colonial Yellow	2094
57	Cream Gold	2884
65	Buckskin	2829
67	Musket Brown Poly	2871
72	Red	2544
78	Firecracker Orange	2084

TWO-TONES: The first two digits are the lower body color, the second two digits the upper color.

1977 BUICK

11	White	2058
13	Silver Poly	2953
15	Gray Poly(Two-Tone)	2862
16	Med. Gray Poly(Two-Tone)	2954
19	Black	9300
22	Light Blue Poly	2955
29	Dark Blue Poly	2959
36	Firethorn Poly	2811
38	Dark Aqua Poly	2961
44	Med.Green Poly	2964
48	Dark Green Poly	2965
51	Yellow	2094

61	Buckskin	2869
63	Gold Poly	2970
69	Brown Poly	2972
72	Red	2973
75	Bright Red	2546
78	Orange Poly	2976
85	Med. Blue Poly(Two-Tone)	2980
91	Blue Firemist Poly	2872
92	Amber Firemist Poly	2873
93	Red Firemist Poly	2987

TWO-TONES: The first two digits are the lower body color, the second two digits the upper color.

1978 BUICK

11	White	2058
15	Silver Poly	3076
16	Gray Poly(Two-Tone)	3077
19	Black	9300
21	Pastel Blue	3078
22	Med. Blue Poly	2955
24	Ult.Mr. Blue Poly	3079
29	Dark Blue Poly	2959
44	Light Green Poly	3081
45	Med. Green Poly	3082
48	Dark Green Poly	2965
51	Yellow	3084
56	Gold Poly(Two-Tone)	3086
61	Tan	3088
63	Dark Gold Poly	3090
67	Saffron Poly	3091
69	Brown	3092
75	Bright Red	3095
77	Red	3096
79	Dark Red	3098
91	Blue Firemist Poly	2872
92	Amber Firemist Poly	2873
93	Red Firemist Poly	2987

TWO-TONES: The first two digits are the lower body color, the second two digits the upper color.

Chapter 3   CADILLAC 1941-1978

PAINT CODE	COLOR	PPG CODE
1941 CADILLAC		
51	Black	9200
52	Antoinette Blue	10072
53	Cavern Green	40075
54	Gunmetal Poly	30052
55	El Centro Green Poly	40086
56	Managua Beige Poly	20046
57	Monica Blue Poly	10086
58	McKinley Gray Poly	30068
59	Valcour Maroon Poly	50021
	Rivermist Gray	30069
	Dusty Gray	30066
	Fairoaks Green Poly	40087
	Berkeley Gray Poly	30070
	Crystal Blue Poly	10087
	Ocean Blue Poly	10088
	Berkshire Green Poly	40074
	Cimarron Green Poly	40123

TWO-TONES

60	U 30069	62	U 30070
64	U 40074		L 30052
	L 30066		
	L 40123		
61	U 40087	63	U 10087
	L 40086		L 10088

PAINT CODE	COLOR	PPG CODE
1942 CADILLAC		
2	Antoinette Blue	10072
3	Cavern Green	40075
4	Gunmetal Gray Poly	30052
5	Ivy Green Poly	40098
6	Pawnee Beige Poly	20211
7	Marlboro Blue Poly	10248
8	Sussex Gray Poly	30269
9	Madeira Maroon Poly	50001
10A	Rockledge Gray	30085
11A	Bahama Blue Poly	10103
14A	Shetland Gray Poly	30086
	Asbury Green Poly	40100
	Ivy Green Poly	40098
	Berkeley Gray Poly	30070
	Devon Green	40099

TWO-TONES

10	U 40099	12	U 40100

14	U 10103		L 40098
	L 30085		
	L 30086		
11	U 30086	13	U 30070
	L 10103		L 30052

1946 CADILLAC

1	Black	9200
2	Antoinette Blue	10072
3	Cavern Green	40075
4	Gunmetal Gray Poly	30052
5	Ivy Green Poly	40098
8	Sussex Gray Poly	30269
9	Madeira Maroon Poly	50001
16	Honey Beige Poly	20270
17	Belden Blue Poly	10304
120	London Gray	30315
121	Seine Blue Poly	10305
	Asbury Green Poly	40100
	Berkeley Gray Poly	30070
	Richmond Gray	30316
	Burbank Green Poly	40350

TWO-TONE

12	U 40100	15	U 30316
20	U 40350		L 30315
	L 40098		
	L 30315		
13	U 30070	19	U 10305
21	U 30315		L 30315
	L 30052		
	L 10305		

1947 CADILLAC

1	Black	9200
2	Antoinette Blue	10072
3	Cavern Green	40075
4	Gunmetal Gray Poly	30052
6	El Paso Beige Poly	20037
9	Maderia Maroon Poly	50001
10	French Gray	30398
11	Seine Blue Poly	10305
14	Camden Green Poly	40446
16	Lucerne Green Poly	10370
17	Belden Blue Poly	10304
18	Dover Gray Poly	30396
	Pinehurst Green Poly	40447
	Vista Gray Poly	30397

22	U 40447	24	U 10370
27	U 30398		L 30398
	L 40446		
	L 10305		
23	U 30052	25	U 30397
	L 30396		L 30398

1948 CADILLAC

1	Black	9200
2	Amherst Blue Poly	10095
3	Cavern Green	40075
4	Tyrolian Gray Poly	30513
5	Cypress Green Poly	40573
6	El Paso Beige Poly	20037
7	Horizon Blue	10472
8	Kingswood Gray Poly	30506
9	Maderia Maroon Poly	50001
10	French Gray	30398
16	Lucerne Green Poly	10370
17	Belden Blue Poly	10304
	Ardsley Green Poly	40572
	Vista Gray Poly	30397

TWO-TONES

12	U 40572	15	U 30397
20	U 10472		L 30398
	L 40573		
	L 10304		
13	U 30506	19	U 30398
	L 30513		L 10370

1949 CADILLAC

1	Black	9200
2	Triumph Blue Poly	10550
3	Dartmouth Green	40692
4	Tyrolian Gray Poly	30513
5	Cypress Green Poly	40573
6	El Paso Beige Poly	20563
7	Horizon Blue	10472
9	Madeira Maroon Poly	50001
10	French Gray	30398
16	Lucerne Green Poly	10370
17	Corinth Blue	10590
18	Avalon Gray Poly	30621
21	Chartreuse	40693
22	Fiesta Ivory	80419
	Ardsley Green Poly	40572
	Kingswood Gray Poly	30506
	Vista Gray Poly	30387

12	U 40572	15	U 30397
20	U 10472		L 30398
	L 40573		
	L 10550		
13	U 30506	19	U 30398
	L 30513		L 10370

1950 CADILLAC

1	Black	9200
2	Hampden Blue Poly	10686
3	Lynton Green Poly	40927
4	Tyrolian Gray Poly	30513
5	Berkshire Blue Poly	10687
6	El Paso Beige Poly	20563
7	Corinth Blue	10590
8	Savoy Gray Poly	30760
9	Maderia Maroon Poly	50001
10	French Gray	30398
12	Glacier Green Poly	40928
22	Fiesta Ivory	80419
	Vista Gray Poly	30397
	Marlow Green Poly	40926
	Kingswood Gray Poly	30506

TWO-TONES

15	U 30397	17	U 30398
20	U 10686		L 10687
	L 30398		
	L 10590		
16	U 40926	19	U 30506
23	U 9200		L 30513
	L 40927		
	L 80419		

1951 CADILLAC

1	Black	9200
2	Empress Blue Poly	10782
3	Exeter Green Poly	41071
4	Capri Green	41072
5	Cadet Blue	10783
6	Tuscon Beige Poly	20779
7	Corinth Blue	10590
8	Savoy Gray Poly	30760
9	Bolero Maroon Poly	50230
10	Mist Gray	30874
12	Chester Green Poly	41073
22	Fiesta Ivory	80419
	Argent Gray Poly	30875

TWO-TONES

15	U 30760	17	U 30875	
20	U 10782		L 10783	
	L 30874			
	L 10590			
16	U 40173	18	U 41071	
23	U 9200		L 41073	
	L 41071			
	L 80419			

1952 CADILLAC

1	Black	9200
2	Empress Blue Poly	10782
3	Inverness Green Poly	41138
4	Alentian Green Poly	41139
5	Nassau Blue Poly	10826
6	Phoenix Beige Poly	20814
7	Olympic Blue	10827
8	Savoy Gray Poly	30760
9	Burgundy Maroon Poly	50286
10	Mist Gray	30874
12	Hillcrest Green Poly	41142
13	Polar Green	41140
14	Opal Gray Poly	31050
22	Sarasota Green	41141

TWO-TONES

15	U 30760	18	U 41138	
23	U 41138		L 41142	
	L 30874			
	L 41141			
16	U 41142	19	U 30760	
	L 41138		L 31050	
17	U 41139	20	U 10826	
	L 41140		L 10827	

1953 CADILLAC

1	Black	9200
2	Cobalt Blue Poly	10967
3	Forest Green Poly	41344
4	Emerald Green Poly	41341
5	Tunis Blue Poly	10964
6	Phoenix Beige Poly	20814
7	Pastoral Blue	10966
8	Norman Gray Poly	30740
9	Burgundy Maroon Poly	50286
10	Court Gray	31085
12	Crystal Green Poly	41342
13	Gloss Green	41340
22	Artisan Ochre	80617
27	Alpine White	31084

28	Azure Blue	10965
29	Aztec Red	70425

TWO-TONE

15	U 30740	17	U 41344	
20	U 10967		L 41340	
	L 31085			
	L 10966			
16	U 41340	18	U 31085	
23	U 9200		L 10964	
	L 41341			
	L 80617			

1954 CADILLAC

1	Black	9200
2	Newport Blue	11149
3	Viking Blue Poly	11150
4	Iris	11151
6	Cobalt Blue Poly	10967
7	Shoal Green	41563
8	Biscay Green	41564
9	Arlington Green Poly	41565
10	Cabot Gray	31187
10Y	Norman Gray Poly	30740
12	Gander Gray Poly	31186
13	Russet Poly	50388
14	Driftwood	21062
14Z	Copper Poly	21063
16	Apollo Gold	80671
17	Aztec Red	70425
18	Alpine White	31084
19	Azure Blue	10965

TWO-TONES

2C	U 11150	7J	U 41565	
14Z	U 21063		L 41563	
	L 11149			
	L 21062			
3S	U 31084	8K	U 31187	
16A	U 9200		L 41564	
	L 11150			
	L 80671			
4S	U 31084	10Y	U 30740	
	L 11151		L 31187	

1955 CADILLAC

10	Black	9200
12	Alabaster Gray	31267
14	Ascot Gray Poly	31260
16	Atlantic Gray Poly	31261
20	Ruskin Blue	11321

22	Azure Blue	10965
24	Dresden Blue Poly	11300
26	Cobalt Blue Poly	10967
30	Mist Green	41729
32	Celadon Green Poly	41743
34	Arlington Green Poly	41565
40	Cape Ivory	80743
42	Pecos Beige	21136
44	Tangier Tan Poly	21145
46	Cocoabar Poly	21138
50	Pacific Coral	60197
52	Mandan Red	70569
54	Deep Cherry Poly	50407
80	Wedgewood Green Light	41730
82	Wedgewood Green Dark Poly	41715
84	Goddess Gold	21146
90	Alpine White	31084
92	Silver Poly	31259

Two Tone combinations can be identified by the paint code shown (patent plate under hood) in four digits. The first two numbers identify the lower color and the last two numbers identify the upper color.

For Example: 3034 U 41565
                      L 41729

1956 CADILLAC

10	Black	9200
11	Canyon Gray	31352
14	Cascade Gray	31353
16	Dawn Gray	31354
18	Camelot Gray Poly	31355
20	Sonic Blue	11475
24	Tahoe Blue	11476
26	Cobalt Blue Poly	10967
30	Duchess Green	41917
32	Princess Green Poly	41918
34	Persian Green	41919
36	Arlington Green Poly	41565
40	Cape Ivory	80743
42	Goddess Gold	21146
44	Pecos Beige	21136
46	Mountain Laurel	70663
48	Taupe Poly	21242
50	Mandan Red	70569
52	Chantilly Poly	50436
90	Alpine White	31084
94	Bahama Blue Poly	11301
96	Emerald Green Poly	41920

1957 CADILLAC

10	Black	9000

12	Alpine White	31084
14	Polo Gray	31460
16	Eton Gray Poly	31461
18	Camelot Gray Poly	31355
20	Orion Blue	11630
24	Tahoe Blue	11476
26	Cobalt Blue Poly	10967
30	Glade Green	42088
32	Thebes Green Poly	42087
34	Turquoise	11631
36	Arlington Green Poly	41565
40	Leghorn Cream	80913
44	Buckskin Beige	21370
46	Mountain Laurel	70748
48	Dusty Rose Poly	70760
48A	Dusty Rose Poly	70788
49	Amethyst Poly	50475
50	Dakotah Red	70757
52	Castile Maroon Poly	50474

1957 CADILLAC BROUGHAM COLORS

110	Ebony	9200
112	Chamonix White	8147
116	Wimbledon Gray Poly	31593
118	Deauville Gray Poly	31592
122	Lake Placid Blue Poly	11711
124	Copenhagen Blue Poly	11709
126	Fairfax Blue Poly	11710
132	Jamaican Green Poly	42180
134	Laurentian Green Poly	42179
136	Plantation Green Poly	42178
140	Manila	80945
144	Sandalwood	21476
148	Kenya Beige Poly	21477
149	Nairobi Pearl Poly	21475
152	Maharani Maroon Poly	50497

ELDORADO, BIARRITZ & SEVILLE

90	Olympic White	8144
92	Starlight Poly	31530
94	Bahama Blue Poly	11301
96	Elysian Green Poly	42109
98	Copper Poly	21417

1958 CADILLAC

10	Black	9000
12	Alpine White	31084
14	Cheviot Gray Poly	31652
16	Prestwick Gray Poly	31654
18	Camelot Gray Poly	31355
20	Daphne Blue	11782

24	Somerset Blue Poly	11781
26	Cobalt Blue Poly	10967
28	Turquoise	11631
29	Peacock Poly	42257
30	Acadian Green	42260
32	Versailles Green Poly	42261
36	Regent Green Poly	42259
40	Calcutta Cream	80976
42	Alamo Beige Poly	21546
44	Buckskin	21370
48	Tahitian Coral Poly	70841
49	Meridian Taupe Poly	50508
50	Dakota Red	70757
90	Olympic White	8144
92	Rajah Silver Poly	31653
94	Argyle Blue Poly	11780
96	Gleneagles Green Poly	42258
98	Desert Gold Poly	21547

1959 CADILLAC

10	Ebony	9300
12	Dover White	8160
14	Silver Poly	31827
16	London Gray Poly	31839
20	Brenton Blue	12003
24	Georgian Blue Poly	12002
26	Dunstan Blue Poly	12016
29	Vegas Turquoise Poly	12001
30	Pinehurst Green	42493
32	Inverness Green Poly	42480
36	Kensington Green Poly	42494
40	Gotham Gold	81090
44	Beaumont Beige	21733
49	Wood Rose Poly	50537
50	Seminole Red	70961
90	Biarritz	8217
	Olympic White	
	Seville	
92	Biarritz	
	Argent Poly	31653
	Seville	
94	Biarritz	
	Argyle Blue Poly	11780
	Seville	
96	Biarritz	
	Hampton Green Poly	42492
	Seville	
98	Biarritz	
	Persian Sand Poly	50541
	Seville	

1960 CADILLAC

10	Ebony Black	9300
12	Olympic White	8217
14	Platinum Gray Poly	31928
16	Aleutian Gray Poly	32017
22	Hampton Blue Poly	12263
24	Pelham Blue Poly	12264
26	York Blue Poly	12224
29	Arroyo Turquoise Poly	12262
32	Inverness Green Poly	42480
36	Glencoe Green Poly	42696
44	Beaumont Beige	21733
45	Palomino	21901
46	Fawn Poly	21722
48	Persian Sand Poly	50541
50	Pompeian Red Poly	50568
94	Lucerne Blue Poly	12265
96	Carrara Green Poly	42725
97	Champagne Poly	21902
98	Siena Rose Poly	71123
99	Heather Poly	50573

In two-tone combinations the first two digits indicate the lower color, the second two digits upper color.

1961 CADILLAC

10	Ebony Black	9300
12	Olympic White	8217
14	Platinum Poly	31928
16	Aleutian Gray Poly	32017
22	Bristol Blue Poly	32099
24	Dresden Blue Poly	12406
26	York Blue Poly	12224
29	San Remo Turquoise Poly	12435
32	Concord Green Poly	42834
34	Lexington Green Poly	42835
36	Granada Green Poly	42836
44	Laredo Tan	22013
46	Tunis Beige Poly	22005
48	Fontana Rose Poly	71202
50	Pompeian Red Poly	50568
94	Nautilus Blue Poly	12407
96	Jade Poly	12405
97	Aspen Gold Poly	22004
98	Topaz Poly	71207
99	Shell Pearl Poly	50595

1962 CADILLAC

10	Ebony	9300
12	Olympic White	8217
14	Nevada Silver Poly	32173
16	Aleutian Gray Poly	32017

22	Newport Blue Poly	12527
24	Avalon Blue Poly	12526
26	York Blue Poly	12224
29	Turquoise Poly	12525
32	Sage Poly	42961
36	Granada Green Poly	42836
44	Sandalwood	22122
45	Maize	81342
46	Driftwood Beige Poly	22121
48	Laurel Poly	71258
50	Pompeian Red Poly	50568
52	Burgundy Poly	50615
94	Neptune Blue Poly	12524
96	Pinehurst Green Poly	42960
97	Victorian Gold Poly	22120
98	Bronze Poly	22119
99	Heather Poly	71257

1963 CADILLAC

10	Ebony	9330
12	Aspen White	8356
14	Nevada Silver Poly	32173
16	Cardiff Gray Poly	32296
22	Benton Blue Poly	12700
24	Basque Blue Poly	12697
26	Somerset Blue Poly	12718
29	Turino Turquoise Poly	12701
32	Basildon Green Poly	43116
36	Brewster Green Poly	43130
44	Bahama Sand	22270
46	Fawn Poly	22273
47	Palomino Poly	22278
48	Briar Rose Poly	71340
50	Matador Red Poly	71339
52	Royal Maroon Poly	50634

In two-tone combinations the first two digits are the lower color, and the last two digits upper color.

ELDORADO COLORS

92	Silver Poly	32299
94	Aquamarine Poly	12703
96	Green Poly	43119
97	Gold Poly	22274
98	Red Poly	71347

1964 CADILLAC

10	Ebony Black	9300
12	Aspen White	8356
14	Nevada Silver Poly	32173
16	Cardiff Gray Poly	32296
22	Beacon Blue Poly	12870

24	Spruce Blue Poly	12869
26	Somerset Blue Poly	12718
29	Turino Turquoise	12701
32	Seacrest Green Poly	43258
34	Lime Poly	43265
36	Nile Green Poly	43259
44	Bahama Sand	22270
46	Sierra Gold Poly	22412
47	Palomino Poly	22278
50	Matador Red Poly	71339
52	Royal Maroon Poly	50634
92	Firemist Blue Poly	12883
94	Firemist Aquamarine Poly	43266
96	Firemist Green Poly	43261
97	Firemist Saddle Poly	22422
98	Firemist Red Poly	71413

In two-tone combinations, the first two digits are the lower color, and the last two digits upper color.

1965 CADILLAC

10	Sable Black	9400
12	Aspen White	8356
16	Starlight Silver Poly	32449
18	Ascot Gray Poly	32477
20	Hampton Blue Poly	13028
24	Tahoe Blue Poly	13029
26	Ensign Blue	13030
28	Alpine Turquoise Poly	43386
30	Cascade Green Poly	43383
36	Inverness Green Poly	43384
40	Cape Ivory	81494
42	Sandalwood	22554
44	Sierra Gold Poly	22412
46	Samoan Bronze Poly	22557
48	Matador Red Poly	71339
49	Claret Maroon Poly	50696
90	Peacock Firemist Poly	13013
92	Sheffield Firemist Poly	32459
96	Jade Firemist Poly	43385
97	Saddle Firemist Poly	22422
98	Crimson Firemist Poly	71413

1966 CADILLAC

10	Sable Black	9400
12	Strathmore White	8631
16	Starlight Silver Poly	32449
18	Summit Gray Poly	32526
20	Mist Blue	13151
24	Marlin Blue Poly	13150
26	Nocturne Blue Poly	13149
28	Caribbean Aqua Poly	43541

30	Cascade Green Poly	43383
36	Inverness Green Poly	43384
40	Cape Ivory	81494
42	Sandalwood	22554
44	Antique Gold Poly	22661
46	Autumn Rust Poly	22663
48	Flamenco Poly	71472
49	Claret Maroon Poly	50696
90	Cobalt Firemist Poly	13158
92	Crystal Firemist Poly	43542
96	Tropic Green Firemist Poly	43500
97	Florentine Gold Firenist Poly	22703
98	Ember Firemist Poly	71542

1967 CADILLAC

10	Sable Black	9400
12	Grecian White	8259
16	Regal Silver Poly	32525
18	Summit Gray Poly	32526
20	Venetian Blue	13359
24	Marina Blue Poly	13361
26	Admiralty Blue Poly	13002
28	Capri Aqua Poly	43662
30	Pinecrest Green Poly	43663
36	Sherwood Green Poly	43390
40	Persian Ivory	81581
42	Sudan Beige	22391
43	Baroque Gold Poly	22814
44	Doeskin Poly	22819
48	Flamenco Red	71472
49	Regent Maroon Poly	50748
90	Atlantis Blue Firemist Poly	13362
92	Crystal Firemist Poly	43666
96	Tropic Green Firemist Poly	43500
97	Olympic Bronze Firemist Poly	71542
98	Ember Firemist Poly	71613

1968 CADILLAC

10	Sable Black	9400
12	Grecian White	8259
16	Regal Silver Poly	32525
18	Summit Gray Poly	32526
20	Arctic Blue	13551
24	Normandy Blue Poly	13553
26	Emperor Blue Poly	13554
28	Caribe Aqua Poly	13552
30	Silverpine Green Poly	43796
36	Ivanhoe Green Poly	43797
40	Kashmir Ivory	81622
42	Sudan Beige	22391
43	Baroque Gold Poly	22996

44	Chestnut Brown Poly	22970
48	San Mateo Red Poly	71642
49	Regent Maroon Poly	50748
90	Spectre Blue Firemist Poly	13555
94	Topaz Gold Firemist Poly	81623
96	Monterey Green Firemist Poly	43798
97	Roosewood Firemist Poly	22972
98	Madeira Plum Firemist Poly	50776

1969 CADILLAC

12	Cotillion White	2058
16	Patina Silver Poly	2059
18	Phantom Gray Poly	2060
24	Astral Blue Poly	2077
26	Athenian Blue Poly	2062
28	Persian Aqua Poly	2063
10	Sable Black	9400
30	Palmetto Green Poly	2080
36	Rampur Green Poly	2065
40	Colonial Yellow	81500
42	Cameo Beige Poly	2066
44	Shalimar Gold Poly	2082
46	Cordovan Poly	2067
47	Wisteria Poly	2068
48	San Mateo Red Poly	71642
49	Empire Maroon Poly	2069
90	Sapphire Blue Firemist Poly	2070
94	Chalice Gold Firemist Poly	2071
96	Biscay Aqua Firemist Poly	2072
97	Nutmeg Brown Firemist Poly	2073
99	Chateau Mauve Firemist Poly	2074

1970 CADILLAC

11	Cotillion White	2058
14	Patina Silver Poly	2059
18	Phantom Gray Poly	2060
24	Corinthian Blue Poly	2164
29	Candor Blue Poly	2167
34	Adriatic Turquoise Poly	2168
42	Lanai Green Poly	2263
19	Sable Black	9400
49	Glenmore Green Poly	2174
54	Byzantine Gold Poly	2177
59	Bayberry Poly	2180
64	Sauterne Poly	2184
69	Dark Walnut	2187
74	San Mateo Red	71642
79	Monarch Burgundy Poly	2191
90	Spartacus Blue Firemist Poly	2192
93	Lucerne Aqua Firemist Poly	2193
94	Regency Gold Firemist Poly	2194

95	Cinnamon Firemist Poly	2195
96	Nottingham Green Firemist Poly	2196
97	Briarwood Firemist Poly	2197
00	Chateau Mauve Firemist Poly	2074

1971 CADILLAC

11	Cotillion White	2058
13	Granoble Silver Poly	2327
16	Oxford Gray Poly	2161
19	Sable Black	9400
24	Zodiac Blue Poly	2328
29	Brittany Blue Poly	2330
34	Adriatic Turquoise Poly	2168
42	Lanai Green Poly	2263
44	Cypress Green Poly	2335
49	Sylvan Green Poly	2337
50	Casablanca Yellow	2175
55	Duchess Gold Poly	2178
64	Desert Beige Poly	2342
69	Clove Poly	2345
74	Cambridge Red Poly	2348
89	Empire Maroon Firemist Poly	2364
90	Bavarian Blue Firemist Poly	2352
92	Pewter Firemist Poly	2353
94	Chalice Gold Firemist Poly	2071
95	Almond Firemist Poly	2354
96	Sausalito Green Firemist Poly	2355
99	Primrose Firemist Poly	2356

1972 CADILLAC

11	Cotillion White	2058
14	Contessa Pewter Poly	2429
18	Mayfair Gray Poly	2430
24	Zodiac Blue Poly	2328
29	Brittany Blue Poly	2330
34	Adriatic Turquoise Poly	2168
44	Sumatra Green Poly	2436
49	Brewster Green Poly	2440
50	Willow	2441
54	Promenade Gold Poly	2442
59	Stratford Covert Poly	2446
64	Tawny Beige Poly	2449
69	Cognac Poly	2452
73	Cambridge Red Poly	2348
74	Coronation Red Poly	2453
90	Ice Blue Firemist Poly	2456
92	St. Moritz Blue Firemist Poly	2457
93	Palomino Firemist Poly	2458
94	Patrician Covert Firemist Poly	2459
96	Balmoral Green Firemist Poly	2460
99	Russet Firemist Poly	2462
19	Sable Black	9400

1973 CADILLAC

11	Cotillion White	2058
13	Georgian Silver Poly	2518
18	Park Avenue Gray Poly	2520
19	Sable Black	9300(9400)
24	Antigua Blue Poly	2523
29	Diplomat Blue Poly	2526
39	Garganey Teal Poly	2527
44	Sage Poly	2529
49	Forest Green Poly	2532
54	Renaissance Gold Poly	2536
63	Laredo Tan Poly	2183
64	Mirage Taupe Poly	2541
68	Burnt Sienna Poly	2543
72	Dynasty Red	2544
81	Harvest Yellow	2549
90	Shadow Taupe Firemist Poly	2550
92	St. Tropez Blue Firemist Poly	2551
94	Phoenix Gold Firemist Poly	2552
95	Oceanic Teal Firemist Poly	2553
96	Viridian Green Firemist Poly	2554
99	Saturn Bronze Firemist Poly	2556

1974 CADILLAC

11	Cotillion White	2058
13	Georgian Silver Poly	2518
18	Deauville Gray Poly	2637
19	Sable Black	9300,9400
24	Antiqua Blue Poly	2523
29	Diplomat Blue Poly	2526
30	Lido Green Poly	2699
34	Mandarin Orange Poly	2701
38	Pueblo Beige Poly	2700
44	Jasper Green	2642
49	Pinehurst Green Poly	2645
54	Promenade Gold Poly	2442
57	Apollo Yellow	2651
59	Canyon Amber	2367
63	Conestoga Tan	2652
69	Chesterfield Brown Poly	2656
71	Andes Copper Poly	2657
72	Dynasty Red	2544
92	Regal Blue Firemist Poly	2660
94	Victorian Amber Firemist Poly	2661
95	Pharaoh Gold Firemist Poly	2662
96	Persian Lime Firemist Poly	2663
98	Terra Cotta Firemist Poly	2664
99	Cranberry Firemist Poly	2665

1975 CADILLAC

| 11 | Cotillion White | 2058 |

28

13	Georgian Silver Poly	2518
15	Vapour Gray	2742
19	Sable Black	9300
24	Jennifer Blue	2745
29	Commodore Blue Poly	2748
30	Lido Green Poly	2699
32	Dunbarton Green	2810
34	Mandarin Orange Poly	2701
36	Firethorn Poly	2811
38	Pueblo Beige Poly	2700
44	Jasper Green	2642
49	Inveraray Green Poly	2752
52	Bombay Yellow	2753
54	Tarragon Gold Poly	2754
65	Knickerbocker Tan	2761
69	Roan Brown Poly	2763
77	Roxena Red	2765
78	Rosewood Poly	2766
92	Gossamer Blue Firemist Poly	2767
94	Galloway Green Firemist Poly	2812
97	Cameo Rosewood Firemist Poly	2768
98	Emberust Firemist Poly	2769
99	Cerise Firemist Poly	2770

1976 CADILLAC

11	Cotillion White	2058
13	Georgian Silver Poly	2518
16	Academy Gray Poly	2862
19	Sable Black	9300
28	Innsbruck Blue Poly	2772
*29	Commodore Blue Poly	2748
*32	Dunbarton Green	2810
*36	Firethorn Poly	2811
37	Claret Poly	2864
38	Pueblo Beige Poly	2700
39	Kingswood Green Poly	2865
50	Calumet Cream	2867
52	Phoenician Ivory	2868
67	Brentwood Brown Poly	2871
69	Chesterfield Brown Poly	2656
90	Crystal Blue Firemist Poly	2872
91	Amberlite Firemist Poly	2873
93	Greenbrier Firemist Poly	2874
*94	Galloway Green Firemist Poly	2812
95	Florentine Gold Firemist Poly	2875
96	Emberglow Firemist Poly	2876

In two-tone combinations the first two digits indicate lower color; the next two digits the upper color.

1977 CADILLAC

| 11 | Cotillion White | 2058 |

13	Georgian Silver Poly	2953
15	Academy Gray Poly (Two-Tone)	2862
19	Sable Black	9300
24	Jennifer Blue	2745
29	Hudson Bay Blue Poly	2959
40	Seamist Green	2962
49	Edinburgh Green Poly	2966
50	Naples Yellow	2884
54	Sovereign Gold Poly	2969
61	Sonora Tan	2869
67	Saffron Poly	2971
69	Demitasse Brn. Poly	2972
74	Bimini Beige	2974
77	Crimson	2975
79	Maderia Maroon	2977
90	Cerulean Blue Firemist Poly	2981
94	Thyme Grn. Firemist Poly	2982
95	Buckskin Firemist Poly	2983
96	Frost Org. Firemist Poly	2984
98	Damson Plum Firemist Poly	2985
99	Desert Rose Firemist Poly	2986

In two-tone combinations the first two digits indicate lower color; the next two digits the upper color.

1978 CADILLAC

11	Cotillion White	2058
15	Academy Gray Poly (Two-Tone)	2862
15	Platinum Poly	3076
16	Pewter Poly	3077
19	Sable Black	9300
21	Columbia Blue	3078
22	Sterling Blue Poly	2955
28	Commodore Blue Poly	2748
40	Seamist Green	2962
49	Blackwatch Green Poly	3083
54	Colonial Yellow	3085
62	Arizona Beige	3089
64	Demitasse Brown Poly	2972
69	Ruidoso Saddle Poly	3092
74	Mulberry Poly	3094
80	Carmine Red	3099
90	Med. Blue Firemist Poly	3100
94	Basil Grn. Firemist Poly	3101
95	Aztec Gold Firemist Poly	3102
96	West. Saddle Firemist Poly	3103
98	Autumn Haze Firemist Poly	3104
99	Canyon Copper Firemist Poly	3105

Chapter 4　　　CHEVROLET 1941-1978

PAINT CODE	COLOR	PPG CODE
1941 CHEVROLET		
288	Ridge Green Poly	40046
289	Banner Beige Poly	20079
290	Ruby Maroon Poly	50014
291	Kingston Gray	30053
292	Cruiser Gray Poly	30108
293	Squadron Gray Poly	30028
294	Constitution Blue Poly	10117
295	Marine Blue Poly	10118
296	Admiral Green Poly	40122
297	Black	9200
298	Cameo Cream	80045
299	Maple Brown Poly	20029
300	Ruby Maroon Poly	50014
	Santone Poly	20081
	Indian Suntan Poly	20028
	Nassak Gray	30029
	Cimarron Green Poly	40123

TWO-TONES

301	U 20081	303	U 40123
	L 20028		L 40046
302	U 30029	304	U 30028
	L 30028		L 10118

1942 CHEVROLET		
308	Maryland Black	9200
309	Martial Maroon Poly	50014
310	Seafoam Green	40064
311	Volunteer Green Poly	40103
312	Sport Beige	20027
313	Santone Beige Poly	20081
314	Torpedo Gray	30053
315	Chevron Gray Poly	30091
316	Ensign Blue Poly	10106
317	Maple Brown Poly	20029
318	Martial Maroon Poly	50014
	Scout Brown	20066
	Fortress Gray	30069
	Wing Blue	10107
	Fleet Blue Poly	10108

TWO-TONES

319	U 40064	321	U 30069
	L 40103		L 30053
320	U 20066	322	U 10107
	L 20027		L 10108

1946 CHEVROLET		
336	Maryland Black	9200
337	Martial Maroon Poly	50014
338	Ensign Blue Poly	10106
	Volunteer Green Poly	40103
	Seafoam Green	40064
	Scout Brown	20066
	Sport Beige	20027
	Fleet Blue Poly	10108
	Wing Gray	10107

TWO-TONES

339	U 40103	340	U 20066	341	U 10108
	L 40064		L 20027		L 10107

1947 CHEVROLET		
344	Black	9200
345	Oxford Maroon Poly	50099
346	Lullwater Green Poly	40451
347	Battleship Gray Poly	30400
348	Sport Beige	20027
349	Freedom Blue	10373
354	Oxford Maroon Poly	50099
355	Maple Brown Poly	20029
	Lakeside Green Poly	40452
	Ozone Blue Poly	10374
	Scout Brown	20066

TWO-TONES

350	U 40451	352	U 10373
	L 20027		L 10374
351	U 40451	353	U 20066
	L 40452		L 20027

1948 CHEVROLET		
347	Battleship Gray Poly	30400
370	Live Oak Green Poly	40566
371	Lake Como Blue Poly	10462
372	Dove Gray Poly	30502
373	Silver Gray Green Poly	40565
375	Oxford Maroon Poly	50099
	Marsh Brown Poly	20455
	Satin Green	40564

TWO-TONES

377	U 20455	379	U 30400
	L 40564		L 30502
378	U 40566	380	U 30502
	L 40565		L 10462

1949 CHEVROLET

Black	9200	
386	Oxford Maroon Poly	50099
387	Monaco Blue Poly	10564
388	Grecian Gray	30628
389	Vista Gray Poly	30629
390	Live Oak Green Poly	40566
391	Satin Green	40564
392	Ice Green Poly	30630
395	Texas Ivory	80404
422	Spruce Green Poly	40899

TWO-TONES

393	U 40566	394	U 30629
	L 40564		L 30628

1950 CHEVROLET

423	Mayland Black	9200
424	Oxford Maroon Poly	50099
425	Grecian Gray	30628
426	Crystal Green Poly	30630
427	Falcon Gray Poly	30742
428	Windsor Blue Poly	10676
429	Mist Green	40912
430	Rodeo Beige Poly	20682
431	Moonlight Cream	80473
438	Empire Red	50124
440	Rodeo Beige Poly	20682

TWO-TONES

432	U 30742	434	U9200	436	U 30742
	L 30628		L 40912		L 80473
433	U 30630	435	U 30628		
	L 40912		L 10676		

1951 CHEVROLET

442	Mayland Black	9200
443	Burgundy Red Poly	50229
444	Thistle Gray	30873
445	Fathom Green Poly	41068
446	Shadow Gray Poly	30872
447	Trophy Blue Poly	10781
447	Regatta Blue Poly	10975
448	Aspen Green Poly	41069
449	Aztec Tan	20776
450	Moonlite Cream	80473
459	Shadow Gray Poly	30889
462	Aspen Green	40569

TWO-TONES

451	U 30872	455	U 30873	460	U 30889
	L 30873		L 30872		L 30873
452	U 41068	456	U 41069	461	U 30873
	L 41069		L 41068		L 30889
453	U 9200	457	U Grained	463	U 41068
	L 80473		L 20776		L 40569
454	U 30873	458	U Grained	464	U 40569
	L 10781		L 41068		L 41068

1952 CHEVROLET

442	Onyx Black	9200
466	Birch Gray	30940
467	Dusk Gray Poly	30941
469	Spring Green	41149
470	Emerald Green Poly	41150
472	Admiral Blue Poly	10831
477	Twilight Blue	10832
478	Sahara Beige	20818
479	Regal Maroon Poly	50289
481	Cherry	70385
482	Honeydew	41148
483	Saddle Brown Poly	20819

TWO-TONES

468	U 30941	487	U 20818
	L 30940		L 50289
471	U 41150	488	U 20819
	L 41149		L 20818
473	U 10831 9200	489	U
	L 10832		L 30940
474	U 41149 9200	491	U
	L 41150		L 50289
484	U 30940	492	U 20820
	L 10832		L 20821
485	U 30940 9200	493	U
	L 41149		L 41148
486	U 20818	494	U 30940
	L 20819		L 10831

1953 CHEVROLET

480	Onyx Black	9200
490	Driftwood Gray	31066
496	Dusk Gray Poly	30941
498	Surf Green	41317
499	Woodland Green Poly	41318
501	Regatta Blue Poly	10781
503	Horizon Blue	10946
504	Sahara Beige	20818
505	Maderia Maroon Poly	50333

506	Target Red	70418
507	Campus Cream	80602
508	Sungold	80612
509	Saddle Brown Poly	20819

TWO-TONES

497	U 30941		513	U 41318
	L 31066			L 80602
500	U 41318		514	U 20819
	L 41317			L 20818
502	U 10781		515	U 20818
	L 10946			L 20819
510	U 80613		516	U 80613
	L 10946			L 80612
511	U 80613		517	U 70418
	L 10781			L 80613
512	U 80602			
	L 41318			

1954 CHEVROLET

540	Onyx Black	9200
541	Surf Green	41317
542	Bermuda Green Poly	41557
543	Horizon Blue	10946
544	Biscayne Blue Poly	11136
545-620	Shoreline Beige	21054
546-619	Saddle Brown Poly	20819
547	India Ivory	80613
548	Shadow Gray Poly	31181
549	Morocco Red	70482
550	Romany Red	70481
551	Fiesta Cream	80668
552	Turquoise	11135
553	Pueblo Tan	21055
618	Sungold	80612

CHEVROLET CORVETTE

567	Polo White	8011
570	Metallic Blue Poly	11238

TWO-TONES

554	U 21054	559	U 21054	564	U 21054
	L 41557		L 20819		L 21055
555	U 80613	560	U 80613	565	U 70482
	L 41317		L 9200		L 21054
556	U 80613	561	U 80613	566	U 21054
	L 10946		L 70481		L 70482
557	U 80613	562	U 41557	621	U 80613
	L 11136		L 80668		L 80612
558	U 41557	563	U 80613	623	U 20819
	L 21054		L 11135		L 21054

1955 CHEVROLET

585	Black	9200
586	Seamist Green	41738
587	Neptune Green Poly	41739
589	Glacier Blue Poly	11329
590	Copper Maroon Poly	50405
591	Shoreline Beige	21054
592	Autumn Bronze Poly	21151
593	India Ivory	8026
594	Shadow Gray Poly	31181
596	Gypsy Red	70575
598	Regal Turquoise Poly	11328
626	Coral	70573
630	Harvest Gold	80739
683	Cashmere Blue	11408
Two-Tones	Navajo Tan	21206
Only	Dusk Rose Poly	50424

TWO-TONES

599	U 41738	608	U 8026	627	U 31181
	L 41739		L 9200		L 70573
600	U 11327	610	U 11329	628	U 9200
	L 11329		L 11327		L 8026
601	U 41739	612	U 8026	629	U 8026
	L 21054		L 11328		L 70573
602	U 8026	613	U 21054	631	U 8026
	L 11327		L 41739		L 80739
603	U 21151	614	U 21054	682	U 8026
	L 21054		L 11329		L 11408
604	U 41739	615	U 21054	684	U 8026
	L 41738		L 70575		L 21206
605	U 8026	616	U 11328	685	U 8026
	L 41738		L 8026		L 50424
606	U 21054	617	U 8026		
	L 21151		L 70575		
607	U 11329	624	U 8026		
	L 21054		L 31181		

1956 CHEVROLET

687	Onyx Black	9200
688	Pinecrest Green	41932
690	Sherwood Green Poly	41933
691	Nassau Blue	11493
692	Harbor Blue Poly	11495
693	Dusk Plum Poly	50440
694	India Ivory	8026
695	Crocus Yellow	80825
697	Matador Red	70673
698	Twilight Turquoise	11494
749	Tropical Turquoise	11586
750	Calypso Cream	80873

752		Inca Silver Poly	31425

Two-Tones Only

		Imperial Ivory	80877
		Grecian Gold Poly	21339
		Laurel Green Poly	41934
		Laurel Green Poly #2	42036
		Dawn Gray Poly	31371
		Adobe Beige	21262
		Dune Beige	21257
		Sierra Gold Poly	21261

TWO-TONES

696	U 9200	707	U 8026	754	U 8026
	L 80825		L 11493		L 11586
700	U 21262	708	U 8026	755	U 80873
	L 21261		L 50440		L 9200
701	U 8026	710	U 8026	756	U 80873
	L 9200		L 11494		L 21339
702	U 41933	711	U 8026	757	U
	L 41932		L 70673		L
703	U 11493	715	U 21257	792	U 80825
	L 11495		L 70673		L 42036
705	U 8026	717	U 80825	763	U 21262
	L 41932		L 41934		L 70673
706	U 8026	721	U 8026		
	L 41933		L 31371		

1957 CHEVROLET

793A	Onyx Black	9000
794A	Imperial Ivory	80877
796A	Harbor Blue Poly	11495
804A		
752	Inca Silver Poly	31425
821A	Dusk Pearl Poly	50472
799A		
749	Tropical Turquoise	11586
823A	Laurel Green Poly	42036
750	Calypso Cream	80873

Two Tones

Only	Grecian Gold Poly	21339
795A	Larkspur Blue	11615
797A	Surf Green	42068
798A	Highland Green Poly	42069
800A	Colonial Cream	80900
801A	Canyon Coral	50471
802A	Matador Red	70673
803A	Coronado Yellow	80901
805A	Sierra Gold Poly	21261
806A	Adobe Beige	21262

Two Tones

Only	India Ivory	8026

TWO-TONE COMBINATIONS

"C" Signifies 2100 Series, Roof & Body to Center Listed on 1st Line Center, Hood, Rear Deck & Body Above Fender on 2nd Line.

807C	8026	812C	42068	817C	8026
	9200		42069		50471
808C	80877	813C	8026	818C	21262
	31425		42068		21261
809C	11495	814C	8026	819C	8026
	11615		80901		70673
810C	8026	815C	80900	820C	80900
	11615		9200		42036
811C	8026	816C	80900	822C	50472
	11586		8026		80877

"D"-2400 Series-Roof Pillar, Body Side Insert Area ON 1st Line Lower Body Listed ON 2nd Line.

807D	8026	812D	42068	817D	8026
	9200		42069		50471
808D	80877	813D	8026	818D	21262
	31425		42068		21261
809D	11615	814D	8026	819D	8026
	11495		80901		70673
810D	8026	815D	9200	820D	80900
	11615		80900		42036
811D	8026	816D	8026	822D	80877
	11586		80900		50472

"E"-1500 Series-Roof & Lower Body Listed ON 1st Line Upper Quarter and Deck Listed ON 2nd Line.

807E	9200	811E	11586	816E	80900
	8026		8026		8026
808E	31425	812E	42068	819E	70673
	80877		42069		8026
809E	11495	813E	42068		
	11615		8026		
810E	11615	815E	80900		
	8026		9200		

1957 CORVETTE

718	Polo White	8011
704	Onyx Black	9200
713	Arctic Blue Poly	11537
709	Aztec Copper Poly	21295
712	Cascade Green	41973
714	Venetian Red	70694

1958 CHEVROLET

900A	Onyx Black	9000
903A	Glen Green	42226

905A	Forest Green Poly	42251
910A	Cashmere Blue	11756
912A	Fathom Blue Poly	11779
914A	Tropic Turquoise	42227
916A	Aegean Turquoise Poly	11778
918A	Anniversary Gold Poly	21526
920A	Sierra Gold Poly	21261
923A	Rio Red	70826
925A	Colonial Cream	80900
930A	Silver Blue Poly	11755
932A	Cay Coral Poly	70840
936A	Snowcrest White	8160

Two-Tones

Only	Arctic White	8161
938A	Honey Beige	21544

"A" Signified Single Tone

TWO-TONE COMBINATIONS
"B" SIGNIFIES 11-1200 SERIES

950B	9200	963B	8161	975B	80900
	8161		42227		8161
953B	8161	964B	11778	980B	70840
	42226		8161		8161
955B	42251	966B	11778	982B	11755
	42226		42227		8160
960B	8161	970B	8161	986B	21526
	11756		21261		21544
962B	11779	973B	70826		
	11756		8161		

"C" SIGNIFIED 15-1600 SERIES

950C	9200	963C	8161	975C	80900
	8161		42227		8161
953C	8161	964C	11778	980C	70840
	42226		8161		8161
955C	42251	966C	11778	982C	11755
	42226		42227		8160
960C	8161	970C	8161	986C	21526
	11756		21261		21544
962C	11779	973C	70826		
	11756		8161		

"D" SIGNIFIES 17-1800 SERIES (EXCEPT 17-1847-67)
(Silver 31751 Wheels, Bumper Center Lower Face Bars.)

950D	9200	963D	8161	975D	80900
	8161		42227		8161
953D	8161	964D	11778	980D	70840
	42226		8161		8161
955D	42251	966D	11778	982D	11755
	42226		42227		8160
960D	8161	970D	8161	986D	21526
	11756		21261		21544
962D	11779	973D	70826		
	11756		8161		

"E" SIGNIFIES CONVERTIBLE 17-1847-67 SERIES
(Special 1847 Sport Coupe)

950E	9200	963E	42227	975E	80900
	8161				8161
953E	42226	964E	11778	980E	70840
			8161		8161
955E	42251	966E	11778	982E	11755
	42226		42227		8160
960E	11756	970E	21261	986E	21544
962E	11779	973E	70826		
	11756		8161		

1958 CORVETTE

510A	Snowcrest White	8160
502A	Silver Blue Poly	11755
504A	Regal Turquoise Poly	11836
	Inca Silver Poly	31425
500A	Charcoal Poly	31742
506A	Signet Red	70826
508A	Panama Yellow	80986

TWO-TONE COMBINATIONS

512B	31742	Inca Silver Poly 31425
	31425	Used on two-tone cove area
514B	11755	and wheels on combination
	31425	numbers 516B, 518B & 520B
516B	11836	
	8160	Snowcrest White 8160
518B	70826	Used on cove area and wheels
	8160	on combination numbers 516B,
520B	80986	518B and 520B.
	8160	
522B	8160	
	31425	

1959 CHEVROLET & CORVETTE

900-A	Tuxedo Black	9300
*503-A		
903-A	Aspen Green Poly	42479
905-A	Highland Green Poly	42495
910-A	Frost Blue Poly	12018
*502-A		
912-A	Harbor Blue Poly	12024
914-A	Crown Sapphire Poly	12001
*504-A		
920-A	Gothic Gold Poly	21723
*509-A	Inca Silver	31425

923-A	Roman Red	70961
*506-A		
925-A	Classic Cream	81092
*508-A		
936-A	Snowcrest White	8160
*510-A		
938-A	Satin Beige	21736
940-A	Grecian Gray Poly	31827
942-A	Cameo Coral Poly	70959

* Corvette Color

TWO-TONE COMBINATIONS

950 - B-C-D-E	8160,	9300
973 - B-C-D-E	8160,	70961
953 - B-C-D-E	8160,	42495
987 - B-C-D-E	12024,	12018
962 - B-C-D-E	12018,	12024
988 - B-C-D-E	8160,	31827
963 - B-C-D-E	8160,	12001
989 - B-C-D-E	21736,	70959
970 - B-C-D-E	21736,	21723
990 - B-C-D-E	42479,	81092

1960 CHEVROLET & CORVAIR

900A		
*902A	Tuxedo Black	9300
^503A		
03A		
*904A	Cascade Green Poly	42693
^517A		
905A		
*906A	Jade Green Poly	42650
910A		
*911A	Horizon Blue Poly	12234
^502A		
912A		
*913A	Royal Blue Poly	12174
915A		
*916A	Tasco Turquoise Poly	12228
^504A		
920A	Suntan Copper Poly	21841
923A		
^506A	Roman Red	70961
925A	Crocus Cream	81202
936A		
*937A	Ermine White	8259
^510A		
938A	Fawn Beige	21873
940A		
*944A	Sateen Silver Poly	31928
941A		

^509A	Shadow Gray Poly	31905

* Corvair Color, ^ Corvette Color

TWO-TONE COMBINATIONS

950	Roof and Pillars	8259
	Lower	9300
963	Roof and Pillars	8259
	Lower	12228
953	Roof and Pillars	8259
	Lower	42693
970	Roof and Pillars	21873
	Lower	21841
955	Roof and Pillars	42693
	Lower	42650
973	Roof and Pillars	8259
	Lower	70961
960	Roof and Pillars	8259
	Lower	12234
984	Roof and Pillars	8259
	Lower	31928
962	Roof and Pillars	12234
	Lower	12174
988	Roof and Pillars	31928
	Lower	31905

1961 CHEVROLET, CORVETTE & CORVAIR

*503	CH,CE,CR	Tuxedo Black	9300
900			
903	CH,CR	Seafoam Green	42838
905	CH,CR	Arbor Green Poly	42837
912	CH,CE,CR	Jewel Blue Poly	12398
*502			
914	CH,CR	Midnight Blue Poly	12397
915	CH,CR	Twilight Turquoise Poly	12396
917	CH,CR	Seamist Turquoise	12401
920	CH,CE,CR	Fawn Beige Poly	22005
*501			
923	CH,CE,CR	Roman Red	70961
*506			
925	5CH,CR	Coronna Cream	81271
936	CH,CE,CR	Ermine White	8259
*510			
938	CH,CR	Almond Beige	21733
940	CH,CE,CR	Sateen Silver Poly	31928
*509			
941	CH	Shadow Gray Poly	31905
948	CH,CE,CR	Honduras Maroon Poly	50568
*523			

* Corvette

CH-Chevrolet, CE-Corvette, CR-Corvair

TWO-TONE COMBINATIONS

950	8259	959	8259	965	12401	973	8259
	9300		12398		12396		70961
953	8259	962	9300	970	21733	984	8259
	42838		12397		22005		31928
955	42838	963	8259				
	42837		12396				

1962 CHEVROLET

900	Tuxedo Black	9300
903	Surf Green	42974
905	Laurel Green Poly	42975
912	Silver Blue Poly	12546
914	Nassau Blue Poly	12552
917	Twilight Turquoise	12550
918	Twilight Blue Poly	12525
920	Autumn Gold Poly	22121
*920	Fawn Beige Poly	22005
923	Roman Red	70961
925	Coronna Cream	81271
927	Anniversary Gold Poly	22157
936	Ermine White	8259
938	Adobe Beige	22137
*938	Almond Beige	21733
940	Satin Silver Poly	32173
*940	Sateen Silver Poly	31928
948	Honduras Maroon Poly	50568

*Corvette

TWO-TONE COMBINATIONS

950	Roof	8259	962	Roof	12546
	Lower	9300		Lower	12552
970	Roof	22137	953	Roof	8259
	Lower	22121		Lower	12525
963	Roof	8259	973	Roof	8259
	Lower	42974		Lower	70961
955	Roof	42974	965	Roof	12550
-	Lower	42975		Lower	12525
984	Roof	8259	959	Roof	8259
	Lower	32173		Lower	12546

1963 CHEVROLET

900	Tuxedo Black	9300
905	Laurel Green Poly	42975
908	Ivy Green Poly	43125
912	Silver Blue Poly	12546
914	Monaco Blue Poly	12711
*916	Daytona Blue Poly	12696
918	Azure Aqua Poly	12525
919	Marine Aqua Poly	43114
920	Autumn Gold Poly	22268

922	Ember Red	71336
*923	Riverside Red	70961
927	Anniversary Gold Poly	22449
932	Saddle Tan Poly	22269
934	Cordovan Brown Poly	22294
936	Ermine White	8259
938	Adobe Beige	22137
940	Satin Silver Poly	32173
*941	Sebring Silver Poly	
948	Palomar Red Poly	50633

* Corvette only

TWO-TONE COMBINATIONS

Body Side Moulding Insert Color is either 8259 Ivory or 8568 Aluminum

950	Roof	8259	963	Roof	8259
	Lower	9300		Lower	12525
972	Roof	22137	954	Roof	8259
	Lower	22294		Lower	42975
967	Roof	12525	973	Roof	8259
	Lower	43114		Lower	71336
959	Roof	8259	970	Roof	22137
	Lower	12546		Lower	22268
984	Roof	8259	971	Roof	22137
	Lower	32173		Lower	22269
962	Roof	12546			
	Lower	12711			

1964 CHEVROLET

900	Tuxedo Black	9300
905	Meadow Green Poly	43264
908	Bahama Green Poly	43263
912	Silver Blue Poly	12546
916	Daytona Blue Poly	12696
918	Azure Aqua Poly	12525
919	Lagoon Aqua Poly	12848
920	Almond Fawn Poly	22392
922	Ember Red	71336
923*	Riverside Red	70961
932	Saddle Tan Poly	22269
936	Ermine White	8259
938	Desert Beige	22391
940	Satin Silver Poly	32173
943	Goldwood Yellow	81450
948	Palomar Red Poly	50633
948	Palomar Red Poly #2	50684

* Corvette only

TWO-TONE COMBINATIONS

952	Roof	43263	965	Roof	8259
988	Roof	12525			

Lower	43264	Lower	12848	Lower	8259
954	Roof 8259	971	Roof 22391		
993	Roof 22391				
Lower	43264	Lower	22269	Lower	50633
959	Roof 8259	975	Roof 22391		
995	Roof 32173				
Lower	12546	Lower	71336	Lower	50633
960	Roof 12696	982	Roof 12696		
Lower	12546	Lower	32173		

1965 CHEVROLET

AA	Tuxedo Black	9300
CC	Ermine White	8259
DD	Mist Blue Poly	13042
EE	Danube Blue Poly	13002
FF	Nassau Blue Poly	13057
GG	Glen Green Poly	43412
HH	Willow Green Poly	43391
JJ	Cypress Green Poly	43390
KK	Artesian Turquoise Poly	43364
LL	Tahitian Turquoise Poly	13003
MM	Milano Maroon Poly	50706
NN	Madeira Maroon Poly	50700
PP	Evening Orchid Poly	50693
QQ	Silver Pearl Poly	32449
RR	Regal Red	71472
SS	Sierra Tan Poly	22553
UU	Rally Red	71491
VV	Cameo Beige	22270
WW	Glacier Gray Poly	32461
XX	Goldwood Yellow	81450
YY	Crocus Yellow	81500

TWO-TONES: The first letter indicates the lower color, the second letter the upper color.

1966 CHEVROLET

AA	Tuxedo Black	9300
CC	Ermine White	8259
DD	Mist Blue Poly	13042
EE	Danube Blue Poly	13002
FF	Marine Blue Poly	13148
HH	Willow Green Poly	43391
KK	Artesian Turquoise Poly	43364
LL	Tropic Turquoise Poly	43496
MM	Aztec Bronze Poly	71525
NN	Madeira Maroon Poly	50700
RR	Regal Red	71472
TT	Sandalwood Tan Poly	22660
VV	Cameo Beige	22270
WW	Chateau Slate Poly	32525
YY	Lemonwood Yellow	81528

TWO-TONES: In two-tone combinations the first letter indicates lower color, the second letter upper color.

1966 CORVETTE

900	Tuxedo Black	9300
972	Ermine White	8259
974	Rally Red	71491
976	Nassau Blue Poly	13057
978	Laguna Blue Poly	13188
980	Trophy Blue Poly	13199
982	Mosport Green Poly	43535
984	Sunfire Yellow	81540
986	Silver Pearl Poly	32449
988	Milano Maroon Poly	50706

1967 CHEVROLET

AA	Tuxedo Black	9300
CC	Ermine White	8259
DD	Nantucket Blue Poly	13349
EE	Deepwater Blue Poly	13346
FF	Marina Blue Poly	13364
GG	Granada Gold Poly	22818
HH	Mountain Green Poly	43651
KK	Ermald Turquoise Poly	43661
LL	Tahoe Turquoise Poly	43659
MM	Royal Plum Poly	50717
NN	Madeira Maroon Poly	50700
RR	Bolero Red	71583
SS	Sierra Fawn Poly	22813
TT	Capri Cream	81578
YY	Butternut Yellow	81500

TWO-TONES: The first letter indicates ower color, the second letter upper color.

CORVETTE

900	Tuxedo Black	9300
972	Ermine White	8259
974	Rally Red	71491
976	Marina Blue Poly	13364
977	Lynndale Blue Poly	13348
980	Elkhart Blue Poly	13347
983	Goodwood Green Poly	43652
984	Sunfire Yellow	81540
986	Silver Pearl Poly	32449
988	Marlboro Maroon Poly	71584

1968 CHEVROLET

AA	Tuxedo Black	9300
CC	Ermine White	8259
DD	Grotto Blue Poly	13512

EE	Fathom Blue Poly	13513
FF	Island Teal Poly	13514
GG	Ash Gold Poly	22942
HH	Grecian Green Poly	43775
JJ	Rallye Green Poly	43898
KK	Tripoli Turquoise Poly	13517
LL	Teal Blue Poly	13516
NN	Cordovan Maroon Poly	50775
PP	Seafrost Green Poly	43774
RR	Matador Red	71634
TT	Palomino Ivory	81617
VV	Sequoia Green Poly	43773
YY	Butternut Yellow	81500

CORVETTE

900	Tuxedo Black	9300
972	Polar White	8631
974	Rally Red	71491
976		
UU	LeMans Blue Poly	13549
978	International Blue Poly	13550
983		
ZZ	British Green Poly	43795
984	Safari Yellow	81621
986	Silverstone Silver Poly	8596
988	Corvette Maroon Poly	50775
992		
OO	Corvette Bronze Poly	22969

TWO-TONES: The first letter indicates lower color, the second letter upper color.

1969 CHEVROLET

10	Tuxedo Black	9300
40	Butternut Yellow	81500
50	Dover White	2058
51	Dusk Blue Poly	2075
52	Garnet Red	2076
53	Glacier Blue Poly	2077
55	Azure Turquoise Poly	2078
57	Fathom Green Poly	2079
59	Frost Green Poly	2080
61	Burnished Brown Poly	2081
63	Champagne Poly	22813
65	Olympic Gold Poly	2082
67	Burgundy Poly	50700
69	Cortez Silver Poly	2059
71	Le Mans Blue Poly	2083
72	Hugger Orange	2084
79	Rallye Green Poly	43898
76	Daytona Yellow	2094

CORVETTE

972	Can-Am White	8631
974	Monza Red	2089
980	Riverside Gold Poly	2092
983	Fathom Green Poly	2079
988	Burgundy Poly	50700
986	Cortez Silver Poly	2059
976	Le Mans Blue Poly	2083
990	Monaco Orange	2084
984	Daytona Yellow	2094
900	Tuxedo Black	9300

TWO-TONES: The first two digits indicate lower color, the next two digits the upper color.

1970 CHEVROLET

10	Classic White	8631
14	Cortez Silver Poly	2059
15	Laguna Gray Poly	2198
17	Shadow Gray Poly	32604
19	Tuxedo Black	9300
25	Astro Blue Poly	2165
26	Mulsanne Blue Poly	2213
27	Bridgehampton Blue Poly	2199
28	Fathom Blue Poly	2166
34	Misty Turquoise Poly	2168
43	Citrus Green Poly	2170
44	Donnybrook Green Poly	2200
45	Green Mist Poly	2171
48	Forest Green Poly	2173
50	Gobi Beige	2175
51	Daytona Yellow	2094
52	Sunflower Yellow	2338
53	Camaro Gold Poly	23211
55	Champagne Gold Poly	2178
58	Autumn Gold Poly	2179
62	Corvette Bronze Poly	2264
63	Desert Sand Poly	2183
65	Hugger Orange	2084
67	Classic Copper Poly	23215
72	Monza Red	2089
75	Cranberry Red	2189
77	Marlboro Maroon Poly	2262
78	Black Cherry Poly	50700

1971 CHEVROLET

11	Antique White	2058
13,905	Nevada Silver Poly	2327
16	Silver Steel Poly	2161
19	Tuxedo Black	9300
24	Ascot Blue Poly	2328

25	Mediterranean Blue	2329
26,976	Mulsanne Blue Poly	2213
29	Command Blue Poly	2330
39	Sea Aqua Poly	2331
42	Cottonwood Green Poly	2333
43	Lime Green Poly	2334
49	Antique Green Poly	2337
52,912	Sunflower	2338
53	Placer Gold Poly	2339
55	Champagne Gold Poly	2178
61	Sandalwood	2181
62	Burnt Orange Poly	2340
63	Mesa Sand	2341
65	Hugger Orange	2084
67	Classic Copper Poly	23215
75	Cranberry Red	2189
78	Rosewood Poly	2350

CORVETTE

10,972	Classic White	8631
27,979	Bridgehampton Blue Poly	2199
48,983	Brands Hatch Green Poly	2336
62,993	Corvette Bronze Poly	2264
76,973	Mille Miglia Red	2349
77,975	Marlboro Maroon Poly	2262
91,989	War Bonnet Yellow Poly	2351
97,987	Ontario Orange Poly	2357
98,988	Steel Cities Gray Firemist Poly	2358

1972 CHEVROLET

11	Antique White	2058
14,924	Pewter Silver Poly	2429
18	Dusk Gray Poly	2430
24	Ascot Blue Poly	2328
25	Mediterranean Blue	2329
26	Mulsanne Blue Poly	2213
28	Fathom Blue Poly	2166
36	Spring Green Poly	2433
43	Gulf Green Poly	2435
46	Oasis Green Poly	2437
48	Sequoia Green Poly	2439
50	Covert Tan	2441
53	Placer Gold Poly	2339
54	Desert Gold Poly	2442
56	Cream Yellow	2444
57	Golden Brown Poly	2445
58	Turin Tan	2463
62	Driftwood	2447
63	Mohave Gold Poly	2448
65	Orange Flame Poly	2450

68	Midnight Bronze Poly	2451
69	Aegean Brown	2452
75	Cranberry Red	2189
19	Tuxedo Black	9300

TWO-TONES: The first two digits indicate lower color, the next two digits the upper color.

CORVETTE

10,972	Classic White	8631
27,979	Targa Blue Poly	2432
37,945	Bryar Blue Poly	2434
47,946	Elkhart Green Poly	2438
52,912	Sunflower Yellow	2338
76,973	Mille Miglia Red	2349
91,989	War Bonnet Yellow Poly	2351
97,987	Ontario Orange Firemist Poly	2357
98,988	Steel Cities Gray Poly	2358

1973 CHEVROLET

11	Antique White	2058
19	Tuxedo Black	9300
23	Medium Blue	2522
24	Light Blue Poly	2523
26	Dark Blue Poly	2524
29	Midnight Blue Poly	2526
41	Medium Green Poly	2437
42	Dark Green Poly	2528
44	Light Green Poly	2529
46	Green-Gold Poly	2530
48	Midnight Green	2531
51	Light Yellow	2533
56	Chamois	2537
60	Light Copper Poly	2538
61	Light Orange	2539
62	Medium Bronze Poly	2540
64	Silver Poly	2541
66	Taupe Poly	2542
68	Dark Brown Poly	2543
74	Dark Red Poly	2545
75	Medium Red	2546
81	Beige	2549
86	Bright Orange	2654
97	Medium Orange	2555

TWO-TONES: The first two digits indicate lower color, the next two digits the upper color.

1973 CORVETTE

10,(910)	Classic White	8631

14,(914)	Corvette Silver Poly	2519
22,(922)	Corvette Med. Blue Poly	2213
27,(927)	Targa Blue Poly	2432
45,(945)	Corvette Bl.-Grn. Poly	2336
47,(947)	Elkhart Green	2438
52,(952)	Corvette Yellow	2534
53,(953)	Corvette Yellow Poly	2535
76,(976)	Mille Miglia Red	2349
80,(980)	Corvette Orange Poly	2548

1974 CHEVROLET

11	Antique White	2058
13	Cosworth Silver Poly	2518
19	Tuxedo Black	9300
24	Light Blue Poly	2523
25	Medium Blue	2639
26	Bright Blue Poly	2524
29	Midnight Blue Poly	2526
36	Aqua Blue Poly	2640
40	Lime-Yellow	2641
44	Medium Green	2642
46	Bright Green Poly	2643
47	Medium Green Poly	2438
49	Medium Dark Green Poly	2645
50	Cream Beige	2646
51	Bright Yellow	2677
53	Light Gold Poly	2649
55	Sandstone	2650
59	Golden Brown Poly	2367
64	Silver Poly	2541
66	Bronze Poly	2653
67	Bright Orange	2654
69	Dark Taupe Poly	2656
74	Medium Red Poly	2658
75	Medium Red	2546
86	Bright Orange	2654

TWO-TONES: The first two digits indicate lower color, the next two digits the upper color.

1974 CORVETTE

10,(910)	Classic White	8631
14,(914)	Corvette Silver Mist Poly	2519
17,(917)	Corvette Gray Poly	2630
22,(922)	Corvette Medium Blue Poly	2213
48,(948)	Dark Green Poly	2644
56,(956)	Corvette Bright Yellow	2094
68,(968)	Dark Brown Poly	2543
74,(974)	Medium Red Poly	2658
76,(976)	Mille Miglia Red	2349
80,(980)	Corvette Orange Poly	2548
13	Silver Poly	2518
74	Red Poly	2658

1975 CHEVROLET

11	Antique White	2058
13	Cosworth Silver Poly	2518
15	Light Gray	2742
16	Medium Gray Poly	2743
19	Tuxedo Black	9300
21	Silver Blue Poly	2431
24	Medium Blue	2745
26	Bright Blue Poly	2746
29	Midnight Blue Poly	2748
44	Medium Green	2642
45	Light Green Poly	2750
49	Dark Green Poly	2752
50	Creme-Beige	2646
51	Bright Yellow	2677
55	Sandstone	2755
58	Dark Sandstone Poly	2757
59	Dark Brown Poly	2758
63	Light Saddle Poly	2759
64	Medium Orange Poly	2760
66	Bronze Poly	2653
72	Medium Red	2544
74	Dark Red	2658
75	Light Red	2546
79	Burgundy Poly	2659
80	Orange Poly	2548

TWO-TONES: The first two digits indicate lower color, the next two digits the upper color.

1975 CORVETTE

10	Classic White	8631
22	Bright Blue Poly	2744
27	Steel Blue Poly	2747
42	Bright Green Poly	2749
56	Bright Yellow	2756
67	Medium Saddle Poly	2762
70	Flame Red	2764
76	Mille Miglia Red	2349

1976 CHEVROLET

11	Antique White	2058
13	Cosworth Silver Poly	2518
16	Medium Gray Poly	2862
19	Tuxedo Black	9300
21,(40)	Light Blue	2815
28	Light Blue Poly	2772
33	Dark Yellow	2814
35,(W35)	Dark Blue Poly	2863
36,(W36)	Firethorn Poly	2811
37,(W37)	Mahogany Poly	2864
40	Lime Poly	2866

45,(47)	Lime Green	2816
49	Dark Green Poly	2752
50	Cream	2867
51	Bright Yellow	2094
57	Cream Gold	2884
65	Buckskin	2829
66	Burnt Orange	2870
67	Medium Saddle Poly	2871
69	Corvette Drk. Brn. Poly	2656
72	Red	2544
75	Red	2546
78	Medium Orange	2084

TWO-TONES: The first two digits indicate lower color, the next two digits the upper color.

1976 CORVETTE

10	Classic White	8631
22	Bright Blue Poly	2744
33	Corvette Drk. Grn. Poly	2877
56	Corvette Bright Yellow	2756
64	Corvette Buckskin	2869
70	Corvette Orange Flame	2764

1977 CHEVROLET

11	Antique White	2058
13	Silver Poly	2953
16	Med. Gray Poly (Two-Tone)	2954
19	Black	9300
21	Light Blue	2815
22	Light Blue Poly	2955
29	Dark Blue Poly	2959
32	Light Lime	2960
36	Firethorn Poly	2811
38	Dark Aqua Poly	2961
44	Med. Grn. Poly	2964
48	Drk. Blue Grn. Poly	2965
50	Cream Gold	2884
51	Bright Yellow	2094
61	Light Buckskin	2869
63	Buckskin Poly	2970
64	Bright Orange	2968
69	Brown Poly	2972
72	Red	2973
75	Light Red	2546
78	Orange Poly	2976
85	Med. Blue Poly (Two-Tone)	2980

TWO-TONES: The first two digits indicate lower color, the next two digits the upper color

1977 CORVETTE

10	Classic White	8631
13	Silver Poly	2953
19	Black	9300
26	Light Blue Poly	2957
28	Dark Blue	2958
52	Yellow	2988
56	Corvette Brt. Yellow	2756
66	Orange	2956
80	Tan Buckskin	2978
83	Dark Red	2979
64	Light Buckskin Poly	

1978 CHEVROLET

11	Antique White	2058
15	Silver Poly	3076
16	Gray Poly (Two-Tone)	3077
19	Black	9300
21	Pastel Blue	3078
22	Light Blue Poly	2955
24	Ultramarine Blue Poly	3079
29	Dark Blue Poly	2959
34	Orange	3070
44	Med. Green Poly	3081
45	Dark Green Poly	3082
48	Dark Blue Green Poly	2965
51	Bright Yellow	3084
56	Gold Poly (Two-Tone)	3086
61	Camel Beige	3088
63	Camel Tan Poly	3090
67	Saffron Poly	3091
69	Dark Camel Poly	3092
75	Red	3095
77	Carmine Poly	3096
79	Dark Carmine Poly	3098

TWO-TONES: The first two digits indicate lower color, the next two digits the upper color

1978 CORVETTE

07	Dark Gray Poly (Two-Tone)	2862
10	Classic White	8631
13	Silver Poly	2953
19	Black	9300
26	Frost Blue	3080
52	Yellow	3072
56	Corvette Bright Yellow	2756
59	Frost Beige	3087
72	Red	2973
82	Mahogany Poly	2864
83	Dark Blue Poly	3074
89	Brown Poly	2656

PAINT CODE	COLOR	PPG CODE
1941 CHRYSLER		
105 & 106	Black	9000
212 & 213	Newport Blue Poly	10050

After Serial Royal 766875, Windsor 7917851
Saratoga 6758411, New Yorker 6631100 use:

222 & 223	Newport Blue	10027
215 & 216	Neutral Blue Poly	10053

After Serial Royal 766875, Windsor 7917851
Saratoga 6758411, New Yorker 6631100 use:

225 & 226	Neutral Blue	10025
223	South Sea Blue	10076
311 & 312	Meadow Green	40003
314 & 315	Polo Green Dk. Poly	40115

After Serial Royal 766875, Windsor 7917851
Saratoga 6758411, New Yorker 6631100 use:

344 & 355	Polo Green Dk.	40000
317 & 318	Porcelain Green Poly	40048

After Serial Royal 766875, Windsor 7917851
Saratoga 6758411, New Yorker 6631100 use:

345 & 346	Polo Green Lt.	40004
350	Spring Gteen	40093
403 & 404	Tropical Tan	20012
511 & 512	Dove Gray	30002
609 & 610	Royal Maroon	50037
611	Sumac Red	70004
812 & 813	Gunmetal	30005
	Skyline Gray Poly	30064
	Skyline Gray	30008

TWO-TONES:

914 & 925	U 30064		929 & 930	U 30064
	L 10053			L 40048
917 & 928	U 50037			
	L 20012			
915 & 926	U 40080		940 & 941	U 30008
	L 40115			L 10025
916 & 927	U 20012		942 & 943	U 40004
	L 50037			L 40000

1942 CHRYSLER

1	Military Blue	10020
2	St. Clair Blue	10003
3	Newport Blue	10027
4	Heather Green	40019
5	Polo Green	40021
6	Meadow Green	40003
7	Dove Gray	30002
8	Gunmetal	30006
9	Catalina Tan	20006
16	Regal Maroon	50037
	Spice Brown	20011

TWO-TONES:

12	U 10020	45	U 40019	78	U 30002
	L 10003		L 40021		L 30006
19	U 20011	54	U 40021	87	U 30006
	L 20006		L 40019		L 30002
21	U 10003	64	U 40003		
	L 10020		L 40019		

1946 CHRYSLER

1	Military Blue	10020
2	St. Clair Blue	10003
3	Newport Blue	10027
4	Heather Green	40019
5	Polo Green	40021
6	Meadow Green	40003
7	Dove Gray	30002
8	Gunmetal	30006
9	Catalina Tan	20006
15	Black	9000
16	Regal Maroon	50037
17	Sumac Red	70004
20	Palace Brick Brown	20124

1947-48 CHRYSLER

9	Catalina Tan	20006
11	Seacrest Green	40523
12	Yellow Lustre	80331
22	Melody Blue Sympho	10338
23	Ballet Taupe Sympho	20305
25	Rossini Brown Sympho	20299
27	Trumpet Gold Sympho	20302
28	Pastorale Green Sympho	40402
29	Andante Green Sympho	40399
30	Noel Green Sympho	40403
32	Pacific Green Sympho	40515
41	Palamino Cream	80323
43	Blue Gray Poly	30455

TWO-TONES:

44	U 20006	46	U 20299
	L 20299		L 20006

1949 CHRYSLER

01	Black	9000
05	Mist Blue	10544
06	Ocean Blue	10566
07	Ensign Blue 10026	
20	Fog Green 40651	
21	Gulf Green 40477	
23	Noel Green Sympho	40403
35	Thunder Gray '	30586
36	Dust Gray 30612	
45	Pearl Tan 20535	
46	Navajo Brown	20466
47	Burmese Brown	20052
60	Burgundy Maroon	50145
61	Pepper Red 70200	
65	Pagoda Cream	80380
66	Anniversary Silver Poly	30631

1950 CHRYSLER

01	Black	9000
05	Haze Blue	10670
06	Racine Blue	10655
07	Newport Blue	10027
20	Fog Green	40651
21	Gulf Green	40477
22	Scotch Green	40687
35	Shell Gray	30543
36	Stone Gray	30734
37	Gunmetal Gray Poly	30735
45	Pearl Tan	20535
46	Tobacco Brown	20592
60	Crown Maroon	50180
61	Victoria Red	50194
65	Pagoda Cream	80469
	Tampa Beige	20688
	Juniper Green Poly	40907
	Indian Brown Poly	20654
	Quaker Gray	30700

TWO-TONES:

70	U 20688	73	U 9000	76	U 20592
	L 9000		L 30700		L 20535
71	U 20688	74	U 9000	77	U 40687
	L 40907		L 30543		L 40651
72	U 20688		U 30735	78	U 10655
	L 20654		L 30734		L 10670

1951-52 CHRYSLER

1	Black	9000
5	Haze Blue	10773
6	Ecuador Blue	10774
7	Newport Blue	10027
20	Foam Green	40984
21	Juniper Green Poly	40907
22	Continental Green Poly	41000
35	Quebec Gray	30876
36	Stone Gray	30734
38	Monitor Gray (1952 only)	30615
45	Arizona Beige	20768
46	Buckskin Tan	20767
47	Indian Brown Poly	20654
60	Crown Maroon	50180
61	Holiday Red	50009
65	Belvidere Ivory	80543

TWO-TONES:

70	U 30876	76	U 20654	82	U 9000
	L 10773		L 20767		L 40984
71	U 30876	77	U 20654	83	U 9000
	L 30734		L 20768		L 40907
72	U 30876	78	U 20767	84	U 30876
	L 40984		L 20768		L 50068
73	U 30876	79	U 41000	85	U 50068
	L 40907		L 40984		L 30876
74	U 30876	80	U 50180	86	U 30876
	L 30735		L 30876		L 30615
75	U 30876	81	U 9000		
	L 50180		L 30876		

1953 CHRYSLER

1	Black	9000
5	Arctic Blue	10918
6	Erie Blue Poly	10916
7	Niagara Blue Poly	10917
8	Columbia Blue Poly	10811
9	Potomac Blue	11105
20	Vermont Green	41251
21	Foliage Green Poly	41252
22	Everglades Green Poly	41253
35	Pearl Gray	31022
37	Submarine Gray Poly	31023
45	Caravan Beige	20877
46	Cinnamon Poly	20885
47	Cocoa Brown Poly	20886
60	Hollywood Maroon Poly	50323
61	Pimento Red	70403
65	Casino Cream	80588

TWO-TONES:

70	U 31022		82	U 41251
	L 10916			9000
71	U 10916		83	U 41253
	L 31022			L 41251
72	U 31022		84	U 41251
	L 31023			L 41253
73	U 31023		85	U 31022
	L 31022			L 9000
74	U 41251		86	U 10918
	L 41252			L 10811
75	U 41252		87	U 10811
	L 41251			L 10918
76	U 20887		88	U 20885
	L 20886			L 20887
77	U 20886		89	U 31022
	L 20887			L 11105
78	U 20887		90	U 11105
	L 20885			L 31022
79	U 20887		91	U 41252
	L 41252			L 80588
80	U 9000		92	U 10811
	L 31022			L 80588
81	U 9000		93	U 11105
	L 41251			L 80588

1954 CHRYSLER

1	Black	9000
5	Alpine Blue	10979
6	Flagship Blue	11105
7	Commodore Blue Poly	10811
8	Glacier Blue	11107
9	Turquoise Blue	11106
10	Peacock Blue(Green) Poly	41528
11	Seabreeze Blue	11221
15	Mint Green	41530
16	Sea Island Green Poly	41529
17	Everglades Green Poly	41253
30	West Point Gray	31144
31	Ascot Gray	31142
40	Pebble Beige	21027
41	Topaz Tan Poly	21026
43	Tahitian Tan	20988
50	Torch Red	70461
55	Canary Yellow	80647
151	Valley Green	41251
301	Canyon Gray	31022
	*Steel Haze Gray	31208
	*Gold	80706
	*Bahama Blue	11211
	*Royal Crest Blue Poly	11217
	*Floral Green	41644
	*Royal Palm Green Poly	41514

*Two-Tones only

TWO-TONES:

60	U 11105		76	U 11107		92	U 9000
	L 31144			L 41528			L 11221
61	U 31144		77	U 41528		93	U 11221
	L 11105			L 11107			L 11217
62	U 10979		78	U 31144		94	U 11217
	L 31144			L 70461			L 11211
63	U 31144		79	U 70461		95	U 41644
	L 10979			L 31144			L 41514
64	U 11105		80	U 41529		96	U 41514
	L 10979			L 80647			L 41644
65	U 10979		81	U 80647		721	U 41251
	L 11105			L 41529			L 41529
66	U 10811		82	U 20988		731	U 41529
	L 10979			L 21028			L 41251
67	U 10979		83	U 21028		611	U 31022
	L 10811			L 20988			L 11105
68	U 31142		84	U 9000		601	U 11105
	L 31144			L 11106			L 31022
69	U 31144		85	U 9000		621	U 10979
	L 31142			L 70461			L 31022
70	U 41253		86	U 70461		631	U 31022
	L 41530			L 9000			L 10979
71	U 4l590		87	U 9000		681	U 31142
	L 41253			L 80647			L 31022
72	U 41530		88	U 80647		691	U 31022
	L 41529			L 9000			L 31142
73	U 41529		89	U 8089		781	U 31022
	L 41530			L 70490			L 70461
74	U 21026		90	U 31208		791	U 70461
	L 21027			L 11221			L 31022
75	U 21027		91	U 31208			
	L 21026			L 80706			

1955 CHRYSLER

1	Black	9000
5	Wisteria Blue	11244
6	Rhapsody Blue Poly	11255
7	Crown Imperial Blue	10278
11	Porcelain Green	41647
12	Shantung Green Poly	41710
13	Jade Green Poly	41650
14	Crown Imperial Green	41671
16	Skyline Gray	31144
17	Embassy Gray Poly	31219
20	Canyon Tan	21117
25	Tango Red	70525

26	Crown Imperial Maroon	50389
27	Navajo Orange	60182
30	Platinum	8096
31	Nugget Gold Poly	20992
32	Sunburst Yellow	80767
	*Desert Sand	21116
	*Falcon Green	41843
	*Heron Blue	11413
	*Two-Tones only	

TWO-TONES:
NEW YORKER SEDAN

35	Body	9000
	Insert	31216
36	Body	11244
	Insert	11255
37	Body	11255
	Insert	11244
38	Body	41647
	Insert	41710
39	Body	41710
	Insert	41647
40	Body	31144
	Insert	31216
41	Body	31216
	Insert	31144

NEW YORKER MODEL C68-I CONV.,
NEWPORT & T & C WAGON

42	Body	21117
	Insert	21116
43	Body	20992
	Insert	8096
44	Body	41650
	Insert	8096
45	Body	11255
	Insert	8096
46	Body	60182
	Insert	21116
47	Body	9000
	Insert	8096
48	Body	8096
	nsert	9000
55	Body	8096
	Roof	9000
Upper front fender	9000	

WINDSOR MODEL

60	Upper	31216
	Lower	9000
61	Upper	11255
	Lower	11244
62	Upper	1244
	Lower	11255
63	Upper	41720
	Lower	41647
64	Upper	41647
	Lower	41710
65	Upper	31216
	Lower	31144
66	Upper	31144
	Lower	31216
67	Upper	9000
	Lower	70525
68	Upper	70525
	Lower	9000
69	Upper	70525
	Lower	70525

CUSTOM IMPERIAL SEDAN

60	Upper	31216
	Lower	9000
61	Upper	11255
	Lower	11244
62	Upper	11244
	Lower	11255
63	Upper	41710
	Lower	41647
64	Upper	41647
	Lower	41710
65	Upper	31216
	Lower	31144
66	Upper	31144
	Lower	31216
58	Upper	8096
	Lower	9000
59	Upper	9000
	Lower	8096

CUSTOM IMPERIAL NEWPORT

67	Upper	9000
	Lower	70525
68	Upper	70525
	Lower	9000
71	Upper	21116
	Lower	21117
72	Upper	21117
	Lower	21116
73	Upper	8096
	Lower	20992
74	Upper	20992
	Lower	8096

75	Upper	8096
	Lower	41650
77	Upper	8096
	Lower	11255
78	Upper	11255
	Lower	8096

NEW YORKER SEDAN

80	Body	9000
	Roof	31216
	Insert	31216
81	Body	11244
	Roof	11255
	nsert	11255
82	Body	11255
	Roof	11244
	Insert	11244
83	Body	41647
	Roof	41710
	Insert	41710
84	Body	41710
	Roof	41647
	Insert	41647
85	Body	31144
	Roof	31216
	Insert	31216
86	Body	31216
	Roof	31144
	Insert	31144

NEW YORKER, C68-1 CONV. NEWPORT & T&C WAGON

87	Body	21117
	Roof	21116
	Insert	21116
88	Body	20992
	Roof	8096
	Insert	8096
89	Body	41650
	Roof	8096
	Insert	8096
90	Body	11255
	Roof	8096
	Insert	8096
91	Body	60182
	Roof	21116
	Insert	21116
92	Body	9000
	Roof	8096
	Insert	8096

NEW YORKER, C-68-2 NEWPORT SPECIAL

93	Body	21116
	Roof	21117
	Hood	21117
	UFF	21117
94	Body	21116
	Roof	60182
	Hood	60182
	UFF	60182
95	Body	8096
	Roof	20992
	Hood	20992
	UFF	20992
96	Body	8096
	Roof	41650
	Hood	41650
97	Body	8096
	Roof	11255
	Hood	11255
	UFF	11255

1955 CHRYSLER SPRING COLOR COMBINATIONS

50	Upper	41650	503	Body	80767
	Lower	80767		Roof	41650
51	Upper	80767		Hood	41650
	Lower	41650		Fender	41650
52	Upper	8096	511	Body	41650
	Lower	80767		Insert	80767
53	Upper	80767	512	Roof	80767
	Lower	8096		Insert	80767
301	Upper	41843		Body	41650
	Insert	8096	521	Body	80767
	Body	41843		Insert	8096
302	Upper	8096	522	Roof	8096
	Insert	8096		Insert	8096
	Body	8096		Body	80767
303	Upper	11413	531	Body	8096
	Insert	8096		Insert	80767
	Body	11413	532	Body	8096
304	Upper	8096		Roof	80767
	Insert	11413		Hood	80767
	Body	8096		Fender	80767
501	Body	80767	533	Roof	80767
	Insert	41650		Insert	80767
502	Roof	41650		Body	8096
	Insert	41650			
	Body	80767			

1956 CHRYSLER

1	Raven Black	9000
5	Stardust Blue	11424

No.	Color	Code
6	Mediterranean Blue Poly	11444
7	Glacier Blue(Green)	41898
8	Turquoise	41899
9	Crown Blue Poly	11242
16	Mint Green	41869
17	Surf Green Poly	41857
18	Hunter Green Poly	41846
19	Crown Green Poly	41858
25	Satin Gray	31324
26	West Point Gray Poly	31226
30	Sand Dune Beige	21185
31	Rosewood Tan	21180
35	Desert Rose	70512
36	Geranium Red	70648
37	Regimental Red	70643
38	Crown Maroon Poly	50428
41	Cloud White	8036
42	Nugget Gold Poly	20992
261	Crocus Yellow	80865
262	Blue Jade	42006
263	Copper Glow Poly	21320

TWO-TONES:

No.		Code	No.		Code	No.		Code	No.		Code
101	R P	8036	104	R P	11424	118	R P	41899	148	R P	70643
	L B	9000		L B	8036		L B	9000		L B	8036
102	R P	9000	105	R P	11424	119	R P	8036	149	R P	9000
	L B	8036		L B	11444		L B	41899		L B	20992
103	R P	8036	106	R P	11444	120	R P	41899	150	R P	20992
	L B	11424		L B	11424		L B	8036		L B	9000
107	R P	31324	137	R P	8036	121	R P	8036	151	R P	8036
	L B	11444		L B	70512		L B	41869		L B	20992
108	R P	11444	138	R P	70512	122	R P	41869	152	R P	20992
	L B	31244		L B	8036		L B	9000		L B	8036
109	R P	8036	139	R P	31226	123	R P	41869	153	R P	31226
	L B	11444		L B	70648		L B	41857		L B	31224
110	R P	11144	140	R P	70648	124	R P	41857	154	R P	31324
	L B	8036		L B	31226		L B	41869		L B	31226
111	R P	9000	141	R P	9000	125	R P	8036	201	R B I	8036
	L B	41898		L B	70648		L B	41857		B	9000
112	R P	41898	142	R P	70648	126	R P	41857	202	R B I	9000
	L B	9000		L B	9000		L B	8036		B	8036
113	R P	31226	143	R P	8036	127	R P	41869	203	R B I	8036
	L B	41898		L B	70648		L B	41846		B	11424
114	R P	41898	144	R P	70648	128	R P	41846	204	R B	111424
	L B	31226		L B	8036		L B	41869		B	8036
115	R P	41899	145	R P	9000	129	R P	8036	205	R B I	11424
	L B	41898		L B	70643		L B	41846		B	11444
116	R P	41898	146	R P	70643	130	R P	41846	206	R B I	11444
	L B	41899		L B	9000		L B	8036		B	11424
117	R P	9000	147	R P	8036	131	R P	21185	207	R B I	31324
	L B	41899		L B	70643		L B	21180		B	11444
						132	R P	21180	208	R B I	11144
							L B	21185		B	31324
						133	R P	8036	209	R B I	8036
							L B	21180		B	11144
						134	R P	21180	210	R B I	11444
							L B	8036		B	8036
						135	R P	9000	211	R B I	9000
							L B	70512		B	41898
						136	R P	70512	212	R B I	41898
							L B	9000		B	41898
						213	R B I	31226	243	R B I	8036
							B	41898		B	70848
						214	R B I	41898	244	R B I	70648
							B	31226		B	8036
						215	R B I	41899	245	R B I	9000
							B	41898		B	70643
						216	R B I	41898	246	R B I	70643
							B	41899		B	9000
						217	R B I	9000	247	R B I	8036
							B	41899		B	70643
						218	R B I	41899	248	R B I	70643
							B	9000		B	8036
						219	R B I	8036	249	R B I	9000

No.		Code	No.		Code
	B	9000		B	2099
220	R B I	41899	250	R B I	20992
	B	8036		B	9000
221	R B I	9000	251	R B I	8036
	B	41869		B	20992
222	R B I	41869	252	R B I	20992
	B	9000		B	8036
223	R B I	41869	253	R B I	31226
	B	41857		B	31324
224	R B I	41857	254	R B I	31324
	B	41869		B	31226
225	R B I	8036	401	A C M	8036
	B	41857		B C M	9000
226	R B I	41857	402	A C M	9000
	B	8036		B C M	8036
227	R B I	41869	403	A C M	8036
	B	41846		B C M	11424
228	R B I	41846	404	A C M	11424
	B	41869		B C M	8036
229	R B I	8036	405	A C M	11424
	B	41846		B C M	11444
230	R B I	41846	406	A C M	11444
	B	41846		B C M	11444
231	R B I	21185	407	A C M	31324
	B	21180		B C M	11444
232	R B I	21180	408	A C M	11444
	B	21185		B C M	31324
233	R B I	8036	409	A C M	8036
	B	21180		B C M	11444
234	R B I	21180	410	A C M	11444
	B	8036		B C M	8036
235	R B I	9000	411	A C M	9000
	B	70512		B C M	41898
236	R B I	70512	412	A C M	41898
	B	9000		B C M	9000
237	R B I	8036	413	A C M	31226
	B	70512		B C M	41998
238	R B I	70512	414	A C M	41998
	B	8036		B C M	31226
239	R B I	31226	415	A C M	41899
	B	70648		B C M	41998
240	R B I	70648	416	A C M	41898
	B	31226		B C M	41899
241	R B I	9000	417	A C M	9000
	B	70648		B C M	41899
242	R B I	70648	418	A C M	41899
	B	41857		B C M	9000
419	A C M	8036	437	A C M	8036
	B C M	41899		B C M	70512
420	A C M	41899	438	A C M	70512
	B C M	8036		B C M	8036

No.		Code	No.		Code
421	A C M	9000	439	A C M	31226
	B C M	41869		B C M	70648
422	A C M	9000	440	A C M	70648
	B C M	9000		B C M	31226
423	A C M	41869	441	A C M	9000
	B C M	41857		B C M	70648
424	A C M	8036	442	A C M	70648
	B C M	41869		B C M	9000
425	A C M	8036	443	A C M	8036
	B C M	41857		B C M	70648
426	A C M	8036	444	A C M	8036
	B C M	41857		B C M	8036
427	C M	8036	445	A C M	9000
	B C M	41846		B C M	70643
428	A C M	41846	446	A C M	70643
	B C M	41869		B C M	9000
429	A C M	8036	447	A C M	8036
	B C M	41849		B C M	70643
430	A C M	41846	448	A C M	70643
	B C M	8036		B C M	8036
431	A C M	21185	449	A C M	9000
	B C M	21180		B C M	20992
432	A C M	21180	450	A C M	20992
	B C M	21185		B C M	9000
433	A C M	8036	451	A C M	8036
	B C M	21180		B C M	20992
434	A C M	21180	452	A C M	20992
	B C M	8036		B C M	8036
435	A C M	9000	453	A C M	31226
	B C M	70512		B C M	31324
436	A C M	70512	454	A C M	31324
	B C M	9000		B C M	31226

THREE-TONES:

No.		Code	No.		Code
301	R	11424	307	R	8036
	A C M	8036		A C M	70512
	B C M	11444		B C M	9000
302	R	41899	308	R	9000
	A C M	41857		A C M	70648
	B C M	41898		B C M	8036
304	R	41846	309	R	9000
	A C M	8036		A C M	8036
	B C M	41857		B C M	70643
305	R	9000	310	R	8036
	A C M	31226		A C M	9000
	B C M	31324		B C M	20992
306	R	9000			
	A C M	21185			
	B C M	21180			

R P   -Roof Panel
L B   -Lower Body

R B I	-Roof & Body Insert	
B	-Body	
A C M	-Above Contour Moulding	
R	-Roof	
B C M	-Below Contour Moulding	

1957 CHRYSLER

A	Jet Black	9000
B	Horizon Blue	11376
C	Regatta Blue Poly	11603
D	Sovereign Blue Poly	11242
E	Seafoam Aqua	42050
F	Parade Green Poly	41826
G	Forest Green Poly	42052
H	Mist Gray	31390
J	Gunmetal Gray Poly	31387
K	Charcoal Gray Poly	31434
L	Desert Beige	21281
M	Shell Pink	21357
N	Copper Brown Poly	21018
P	Gauguin Red	70693
R	Regimental Red	70643
S	Sunset Rose	70696
T	Champagne Gold	80898
U	Deep Ruby Poly	50467
V	Saturn Blue	11527
W	Indian Turquoise	42006
	*262 Blue Jade	
X	Cloud White	8036
	*261 Crocus Yellow	80865
	*263 Copper Glow Poly	21320

*1956 Spring colors, two-tones only

TWO-TONES and **TWO-TONES WITH INSERT**: First letter indicates upper color, second letter lower color and third letter insert color.

1958 CHRYSLER

AAA	Raven Black	9000
BBB	Stardust Blue	11684
CCC	Air Force Blue Poly	11596
DDD	Midnight Blue	11679
EEE	Spring Green	41415
FFF	Cypress Green Poly	41494
GGG	Mandarin Jade Poly	42149
HHH	Aztec Turquoise	42150
JJJ	Spruce Green Poly	42139
KKK	Satin Gray	31552
LLL	Winchester Gray Poly	31544
MMM	Mesa Tan	21447
NNN	Sandalwood Poly	21364

OOO	Tahitian Coral	70749
PPP	Matador Red	70791
RRR	Shell Pink	21357
TTT	Garnet Maroon Poly	50483
UUU	Bamboo Yellow	80062
WWW	Ballet Blue	11690
XXX	Ermine	8131
ZZZ	Champagne Gold	80898

TWO-TONES and **TWO-TONES WITH INSERT:** First letter indicates upper color, second letter lower color and third letter insert color.

1959 CHRYSLER

AAA	Formal Black	9000
BBB	Normandy Blue	11697
CCC	Nocturne Blue Poly	11786
DDD	Empress Blue Poly	11794
EEE	Ballad Green	42172
FFF	Highland Green Poly	42262
GGG	Poly	42253*
HHH	Silverpine Poly	42264
III	Tropic Turquoise	12170
JJJ	Aqua Mist	11787
KKK	Turquoise Gray Poly	42263
LLL	Spanish Silver Poly	31539
MMM	Oxford Gray Poly	31660*
MMM	Storm Gray Poly	31663
NNN	Persian Pink	70848
PPP	Carousel Red	70911
RRR	Radiant Red	70791
SSS	Gray Rose Poly	50509
SSS	Fireglow Polyx 60314(1958 Spring Color)	
TTT	Deep Ruby Poly	50510
UUU	Sandstone	21529
WWW	Cameo Tan Poly	21551
XXX	Ivory White	8131
YYY	Yellow Mist	80062*
YYY	Spun Yellow	80980
ZZZ	Copper Spice Poly	21550
VVV	Bimini Blue 11922(1958 Spring Color)	
YYY	Frosty Tan Poly 21487(1958 Spring Color)	

*Imperial only

TWO-TONES and **TWO-TONES WITH INSERT:** First letter indicates upper color, second letter lower color and third letter insert color.

1960 **CHRYSLER**

AA-1	Sunburst	81125
BB-1	Formal Black	9000
CC-1	Starlight Blue	11988(C)
CC-1	Glacier Blue	12047(I)
DD-1	Polar Blue Poly	12044(C)
DD-1	Moonstone Blue Poly	12048(I)
EE-1	Midnight Blue Poly	12049(I)
FF-1	Surf Green	42539(C)
FF-1	Light Mint	41342(I)
GG-1	Ivy Green Poly	42538(C)
GG-1	Cedar Green Poly	42542(I)
HH-1	Silverpine Poly	42746
JJ-1	Seaspray	11787(C)
KK-1	Bluegrass Poly	42263(C)
LL-1	Sheffield Silver Poly	31746
NN-1	Executive Gray Poly	31660
OO-1	Regent Ruby	50546(I)
PP-1	Toreador Red Poly	71003(C)
PP-1	Regal Red	71030(I)
RR-1	Dawn Mauve	50544(I)
RR-1	Lilac	12045(C)
SS-1	Iris Poly	12046(C)
SS-1	Dusk Mauve Poly	50545(I)
TT-1	Daytona Sand	21758(C)
TT-1	Beach Beige	21761(I)
UU-1	Autumn Haze Poly	21759(C)
UU-1	Powdered Bronze Poly	21765(I)
WW-1	Alaskan White	8218
YY-1	Petal Pink	71004
ZZ-1	Terra Cotta Poly	71053
	C) Chrysler only	
	(I) Imperial only	

TWO-TONES:

WINDSOR & SARATOGA: eg.WA-2: First letter is roof and insert color, second is basic car color.
NEW YORKER: eg. WA-2: First letter is roof color, second is basic car color.
IMPERIAL CUSTOM & CROWN: eg. WA-2: First letter is roof color, second is basic car color.
IMPERIAL LE BARON: eg. WA-2: First letter is steel insert color, second is basic car color.

1961 CHRYSLER & IMPERIAL

AA-1	Coronado Cream	21925(I)
BB-1	Formal Black	9000
CC-1	Parisian Blue	12216(C)
CC-1	Ice Blue	12319(I)
DD-1	Capri Blue Poly	12272
EE-1	Midnight Blue Poly	12294(I)

GG-1	Pinehurst Green Poly	42732
JJ-1	Tahitian Turquoise	11273(C)
KK-1	Teal Blue Poly	42773(I)
LL-1	Sheffield Silver Poly	31746
MM-1	Dove Gray	32127
NN-1	Executive Gray Poly	32100(I)
OO-1	Dubonnet Poly	71131(C)
PP-1	Mardi Gras Red	71203(C)
PP-1	Coronation Red	71136(I)
RR-1	Cinnamon Poly	71140(C)
WW-1	Alaskan White	8218
YY-1	Sahara Sand	21905(C)
YY-1	Malibu Tan	21934(I)
ZZ-1	Tuscan Bronze Poly	21927(C)
ZZ-1	Autumn Russet Poly	21921(I)
	(C) Chrysler only	
	(I) Imperial only	

TWO-TONES:

NEWPORT & WINDSOR: eg. WA-2: First letter is roof and insert color, second is basic car color.
NEW YORKER: eg. WA-2: First letter is roof color, second is basic car color.
IMPERIAL CUSTOM & CROWN: eg. WA-2: First letter is insert color, second is basic car color.
IMPERIAL LE BARON: eg. WA-2: First letter is steel insert color, second is basic car color.

1962 CHRYSLER

BB-1	Formal Black	9000
CC-1	Dawn Blue	12403
DD-1	Sapphire Blue Poly	12416
EE-1	Moonlight Blue Poly	12415(I)
FF-1	Willow Green	42840
GG-1	Sage Green Poly	42854
JJ-1	Bermuda Turquoise	12414(C)
LL-1	Limelight	43030(C)
MM-1	Dove Gray	32127
NN-1	Alabaster	32108(I)
OO-1	Embassy Red	50599(I)
PP-1	Festival Red	71203(C)
RR-1	Silver Lilac Poly	21966(I))
SS-1	Cordovan Poly	22025(I)
TT-1	Coral Gray	32074(C)
VV-1	Seascape	43051(C)
WW-1	Oyster White	8293
YY-1	Rosewood Poly	22023
ZZ-1	Caramel	22095
	(C) Chrysler only	
	(I) Imperial only	

TWO-TONES: eg BW-2: First letter is roof color, second letter is body color.

1963 CHRYSLER

BB-1	Formal Black	9000
CC-1	Glacier Blue	12469
DD-1	Cord Blue Poly	12721
EE-1	Navy Blue Poly	12722
GG-1	Surf Green Poly	43131(C)
HH-1	Forest Green Poly	43134
KK-1	Holiday Turquoise Poly	12724
LL-1	Teal Poly	12723
MM-1	Alabaster	32202
NN-1	Madison Gray Poly	32306
OO-1	Charcoal Poly	32306(I)
PP-1	Festival Red	71203(C)
RR-1	Mayan Gold Poly	22284(I)
SS-1	Ivory	81355(I)
TT-1	Claret Poly	71348
VV-1	Pace Car Blue Poly	12831(C)
UU-1	Embassy Gold Poly	22311(C)
WW-1	Oyster White	8293
XX-1	Fawn	22234
YY-1	Cypress Tan Poly	22282(C)
ZZ-1	Mahogany Poly	22285(I)

(C) Chrysler only
(I) Imperial only

TWO-TONES: eg BW-2: First letter is roof color, second letter is body color.

1964 CHRYSLER

BB-1	Formal Black	9000
CC-1	Wedgewood	12655
DD-1	Nassau Blue Poly	12763
EE-1	Monarch Blue Poly	12764
FF-1	Pine Mist Poly	43151
GG-1	Seguoia Green Poly	43149
KK-1	Silver Turquoise Poly	12648
LL-1	Royal Turquoise Poly	12765
MM-1	Madison Gray Poly	32305
NN-1	Charcoal Gray Poly	32306(I)
OO-1	Rosewood Poly	50635
RR-1	Royal Ruby Poly	50638
SS-1	Ivory	81403(I)
TT-1	Roman Red Poly	71393
UU-1	Embassy Gold Poly	22311(C)
WW-1	Persian White	8358
XX-1	Dune Beige	22293
YY-1	Sable Tan Poly	22317
22-1	Silver Mist Poly	32398(C)

22-9	Silver Mist Poly	32398(C)

(C) Chrysler only
(I) Imperial only

TWO-TONES: eg BW-2: First letter is roof color, second letter is body color.

1965 CHRYSLER

AA-1	Regal Gold Poly	22461
BB-1	Formal Black	9000,9900
CC-1	Ice Blue	12894
DD-1	Nassau Blue Poly	12763
EE-1	Navy Blue Poly	12896
FF-1	Mist Blue Poly	12895
GG-1	Sequoia Green Poly	43149
KK-1	Peacock Turquoise Poly	12897
LL-1	Royal Turquoise Poly	12765
MM-1	Granite Gray Poly	32401
NN-1	Silver Mist Poly	32398
RR-1	Sierra Sand	22441
SS-1	French Ivory	81413
TT-1	Spanish Red Poly	71476
VV-1	Cordovan Poly	50673
WW-1	Persian White	8362
XX-1	Sand Dune Beige	22440
YY-1	Sable Tan Poly	22643
ZZ-1	Frost Turquoise Poly	12898
22-1	Sage Green Poly	43287
33-1	Pink Silver Poly	22444
44-1	Moss Gold Poly	22443
55-1	Black Plum Poly	50672
66-1	Mauve Poly	50771
77-1	Patrician Gold Poly	22442

TWO-TONES: First letter or number is roof color, second letter or number is body color.

1966 CHRYSLER

AA-1	Silver Mist Poly	32398
BB-1	Formal Black	9000,9300
CC-1	Powder Blue	13037
DD-1	Crystal Blue Poly	13043
EE-1	Regal Blue Poly	13040
FF-1	Haze Green Poly	43414
GG-1	Sequoia Green Poly	43419
KK-1	Frost Turquoise Poly	12898
LL-1	Royal Turquoise Poly	12765
PP-1	Scorch Red	71483
QQ-1	Spanish Red Poly	71476
RR-1	Daffodil Yellow	81515
SS-1	Ivory	81501

WW-1	Persian White	8362
XX-1	Desert Beige	22541
YY-1	Saddle Bronre Poly	22538
ZZ-1	Spice Gold Poly	22511
33-1	Dove Tan	22542
44-1	Moss Gold Poly	22443
55-1	Dusty Gold Poly	22560
66-1	Lilac Poly	50702
77-1	Ruby Poly	50699
88-1	Deep Plum Poly	50701
88-1	Daffodil Yellow	81515

TWO-TONES: First letter or number is roof color, second letter or number is body color.

1967 CHRYSLER

AA-1	Silver Mist Poly	32398
BB-1	Formal Black	9000,9300
CC-1	Arctic Blue Poly	13159(C)
CC-1	Aegean Blue Poly	13159(I)
DD-1	Crystal Blue Poly	13043(C)
DD-1	Wedgewood Blue Poly	13043(I)
EE-1	Regal Blue Poly	13040
FF-1	Mint Green Poly	43547(C)
FF-1	Haze Green Poly	43547(I)
GG-1	Pine Green Poly	43540(C)
GG-1	Forest Green Poly	43540(I)
JJ-1	Mahogany Poly	21704(C)
JJ-1	Sepia Poly	22704(I)
KK-1	Mist Turquoise Poly	13135(C)
KK-1	Aqua Turquoise Poly	13135(I)
LL-1	Twilight Turquoise Poly	13214
MM-1	Turbine Bronze Poly	60492
PP-1	Scorch Red	71483(C)
PP-1	Flame Red	71483(I)
QQ-I	Ruby Red Poly	71552(C)
QQ-1	Plum Red Poly	71552(I)
RR-1	Daffodil Yellow	81515
SS-1	Ivory	81501
WW-1	Persian White	8362
XX-1	Sandalwood	22701(C)
XX-1	Imperial Navaho Beige	22701(I)
YY-1	Desert Dune Poly	22700(C)
YY-1	Imperial Fawn Poly	22700(I)
ZZ-1	Spice Gold Poly	22715(C)
ZZ-1	Cinnamon Gold Poly	22715(I)
55-1	Charcoal Gray Poly	32599
66-1	Mauve Mist Poly	50731(C)
66-1	Dusty Pink Poly	50731(I)
77-1	Ruby Poly	50699(I)
88-1	Mediterannean Blue Poly	13336(C)

(C) Chrysler only
(I) Imperial only

TWO-TONES: eg BW-2: First letter or number is roof color, second letter or number is body color.

1968 CHRYSLER

AA-1	Silver Haze	8588
BB-1	Formal Black	9000,9300
CC-1	Consort Blue Poly	13355
DD-1	Sky Blue Poly	13360
EE-1	Military Blue Poly	13372
FF-1	Frost Green Poly	43646
GG-1	Forest Green Polv	43649
HH-1	Antique Ivory	81575(C)
HH-1	Champagne	81575(I)
JJ-1	Sovereign Gold Poly	22807
KK-1	Mist Turquoise Poly	13195
MM-1	Turbine Bronze Poly	60492
PP-1	Scorch Red	71483(C)
PP-1	Flame Red	71483(I)
RR-1	Burgundy Poly	50749
TT-1	Meadow Green Poly	49647
WW-1	Polar Whitye	8653
XX-1	Sandalwood	22441(C)
XX-1	Imperial Navaho Beige	22441(I)
YY-1	Beige Mist Poly	22855
44-1	Bright Turquoise Poly	13534
55-1	Charcoal Gray Poly	32599
999	Special Order Colors	
	Corporate Blue	8367
	Corporate White	12785

(C) Chrysler only
(I) Imperial only

TWO-TONES: eg BW-2: First letter or number is accent or roof color, second letter or number is basic body color.

1969 CHRYSLER

A-4	Platinum Poly	2016
A-9	Charcoal Poly	2017(I)
A-9	Dark Gray Poly	2017(C)
B-3	Bahama Blue	2018
B-7	Jubilee Blue Poly	2020(C)
B-9	Midnight Blue Poly	2021(I)
E-7	Dark Briar Poly	2022
F-3	Surf Green Poly	2023
F-5	Avocado Poly	2024(C)
F-8	Jade Green Poly	49786
F-9	Dark Emerald Poly	2026(I)

L-1	Sandalwood	22542(C)
L-1	Navaho Beige	22541(I)
M-9	Deep Plum	2027
Q-4	Aquamarine Poly	2028
R-6	Crimson	2029(C)
T-3	Bronze Mist Poly	2030
T-5	Burnished Bronze Poly	2031(C)
T-7	Tuscan Bronze Poly	2032
W-1	Spinnaker White	2033
Y-3	Antique Ivory	81575(C)
Y-3	Champagne	81575(I)
Y-4	Classic Gold Poly	2034
Y-5	Mystic Gold Poly	2117
X-9	Formal Black	9300
999	Special Order Colors	
	(C) Chrysler only	
	(I) Imperial only	

TWO-TONES: First two digits are accent or roof colors, second two digits are basic body color.

1970 CHRYSLER

A-4	Platinum Poly	2016
A-9	Charcoal Poly	2017
B-3	Bahama Blue Poly	2018
B-7	Jubilee Blue Poly	2020
F-4	Lime Green Poly	2133
F-8	Jade Green Poly	43786
F-9	Dark Emerald Poly	2026
L-1	Sandalwood	22542(C)
L-1	Navaho Beige	22542(I)
L-6	Aztec Gold Poly	2261
M-9	Deep Plum	2027
P-6	Teal Poly	2132
R-6	Crimson	2029(C)
R-8	Burgundy Poly	50749
T-3	Satin Tan Poly	2131
T-6	Deep Bronze Poly	2129
T-8	Walnut Poly	2130(I)
W-1	Spinnaker White	2033
Y-3	Antique Ivory	81575(C)
Y-3	Champagne	81575(I)
Y-4	Mystic Gold Poly	2117
Y-6	Citron Gold Poly	2102
X-9	Formal Black	9300
999	Special Order Colors	
	(C) Chrysler only	
	(I) Imperial only	

TWO-TONES: First two digits are accent or roof colors, second two digits are basic body color.

1971 CHRYSLER

A-4	Winchester Gray Poly	2314
A-8	Slate Gray	2315
A-9	Charcoal Poly	2017(I)
B-2	Glacial Blue Poly	2304
B-7	Evening Blue Poly	2302
B-7	Midnight Blue Poly	2302(C)
E-5	Rallye Red	2136(C)
E-7	Burnished Red Poly	2321
F-3	Amber Sherwood Poly	2316
F-9	Avocado Poly	2318
J-4	April Green Poly	2319
K-6	Autumn Bronze Poly	2312
L-1	Sandalwood Beige	22542
L-6	Aztec Gold Poly	2261
M-8	Sparkling Burgundy Poly	2322(I)
Q-5	Coral Turquoise Poly	2301
T-8	Tahitian Walnut Poly	2309
W-1	Spinnaker White	2033
X-9	Formal Black	9300
Y-1	Lemon Twist	2211(C)
Y-4	Honeydew	2310(I)
Y-9	Tawny Gold Poly	2311
999	Special Order Colors	
	(C) Chrysler only	
	(I) Imperial only	

TWO-TONES: First two digits are accent or roof colors, second two digits are basic body color.

1972 CHRYSLER

A-5	Silver Frost Poly	2513
A-9	Charcoal Poly	2017
B-1	Blue Sky	2424
B-5	True Blue Poly	2306
B-7	Evening Blue Poly	2302
B-9	Regal Blue POly	2508
E-5	Red	2136
E-7	Burnished Red Poly	2321
F-1	Mist Green	2515
F-3	Amber Sherwood Poly	2316
F-7	Sherwood Green Poly	2317
F-8	Forest Green Poly	2514
JY-9	Tahitian Gold Poly	2510
K-6	Autumn Bronze Poly	2312
L-4	Sahara Beige	2427
Q-5	Coral Turquoise Poly	2301
T-8	Chestnut Poly	2425
W-1	Spinnaker White	2033
Y-2	Sun Fire Yellow	81574
Y-3	Honey Gold	2517

53

Y-4	Honeydew	2310
Y-6	Gold Leaf Poly	2307
Y-9	Tawney Gold Poly	2311
X-9	Formal Black	9300
999	Special Order Colors	

TWO-TONES: First two digits are accent or roof colors, second two digits are basic body color.

1973 CHRYSLER

A-5	Silver Frost Poly	2513
B-1	Blue Sky	2424
B-5	True Blue Polv	2306
B-9	Regal Blue Poly	2508
E-7	Burnished Red Polv	2321
F-1	Mist Green	2515
F-3	Amber Sherwood	2316
F-8	Forest Green Poly	2514
K-3	Navaho Copper Poly	2586
L-4	Sahaia Beige	2457
Q-5	Coral Turquoise Poly	2301
W-1	Spinnaker White	2033
X-9	Formal Black	9000,9300
Y-2	Sun Fire Yellow	81574
Y-3	Honey Gold	2517
Y-6	Golden Haze Poly	2509
JY-9	Tahitian Gold Poly	2510
999	Special Order Colors	

TWO-TONES: First two digits are accent or roof colors, second two digits are basic body color.

1974 CHRYSLER

A-5	Silver Frost Poly	2513
B-1	Powder Blue	2626
B-5	Lucerne Blue Poly	2627
B-8	Starlight Blue Poly	2628
E-7	Burnished Red Poly	2321
G-2	Frosty Green Poly	2629
G-8	Deep Sherwood Poly	2631
J-6	Avocado Gold Poly	2632
K-3	Navaho Copper Poly	2586
L-4	Sahara Beige	2427
L-8	Dark Moonstone Poly	2633
T-5	Sienna Poly	2634
T-9	Dark Chestnut Poly	2590
W-1	Spinnaker White	2033
X-9	Formal Black	9300
Y-2	Sunfire Yellow	81574
Y-4	Golden Fawn	2635
Y-9	Tahitian Gold Poly	2510
999	Special Order Colors	

TWO-TONES: First two digits are accent or roof colors, second two digits are basic body color.

1975 CHRYSLER

A-2	Silver Cloud Poly	2734
B-1	Powder Blue	2626
B-2	Astral Blue Poly	2735
B-5	Lucerne Blue Poly	2627*
B-8	Starlight Blue Poly	2628
E-5	Rallye Red	2136*
E-9	Vintage Red Poly	2736
G-2	Frosty Green Poly	2629
G-8	Deep Sherwood Poly	2631
J-2	Platinum Poly	2730
J-6	Avocado Gold Poly	2632
K-3	Bittersweet Poly	2740
L-4	Sahara Beige	2427
L-5	Moondust Poly	2737
T-4	Cinnamon Poly	2741*
T-5	Sienna Poly	2634
T-9	Dark Chestnut Poly	2590
W-1	Spinnaker White	2033
X-9	Formal Black	9300
Y-4	Golden Fawn	2635
Y-5	Yellow Blaze	2636*
Y-6	Inca Gold Poly	2738
Y-9	Spanish Gold Poly	2739
	*Cordoba only	
999	Special Order Colors	

TWO-TONES: First two digits are accent or roof colors, second two digits are basic body color.

1976 CHRYSLER

A-1	Brite Silver Poly	2888
A-2	Silver Cloud Poly	2734
A-5	Silver Frost Poly	2513
B-1	Powder Blue	2626
B-2	Astral Blue Poly	2735
B-5	Jamaican Blue Poly	2851
B-8	Starlight Blue Poly	2628
E-5	Rallye Red	2136
E-8	Vintage Red Sunfire Poly	2849
E-9	Vintage Red Poly	2736
F-2	Jade Green Poly	2852
G-8	Deep Sherwood Poly	2631
G-9	Deep Sherwood Sunfire Poly	2848
J-2	Platinum Poly	2730
J-5	TropiC Green Poly	2853
K-3	Bittersweet Poly	2740
L-4	Sahara Beige	2427

L-5	Moondust Poly	2737
T-9	Dark Chestnut Poly	2590
U-2	Saddle Tan	2855
U-3	Carmel Tan Poly	2856
U-6	Light Chestnut Poly	2857
W-1	Spinnaker White	2033
X-9	Formal Black	9000,9300
Y-1	Jasmine Yellow	2946
Y-4	Golden Fawn	2635
Y-6	Inca Gold Poly	2738
Y-7	Taxi Yellow	81746
Y-9	Spanish Gold Poly	2739
999	Special Order Colors	

TWO-TONES: First two digits are accent or roof colors, second two digits are basic body color.

1977 CHRYSLER

A-1	Burnished Silver Poly	2888
A-2	Silver Cloud Poly	2734
B-2	Wedgewood Blue	2934
B-3	Cadet Blue Poly	2935
B-9	Starlt. Blue Sunfire Poly	2938
E-8	Vintage Red Sunfire Poly	2849
F-2	Jade Green Poly	2852
F-7	Forest Green Sunfire Poly	2939
K-6	Burnished Copper Poly	2940
L-3	Mojave Beige	2941
L-5	Moondust Poly	2737
R-6	Claret Red	2854
R-8	Russet "Sunfire" Poly	2942
T-2	Light Mocha Tan	2943
T-7	Coffee Sunfire Poly	2944
U-3	Caramel Tan Poly	2856
U-6	Light Chestnut Poly	2857
W-1	Spinnaker White	2033
X-8	Forest Black Sunfire Poly	2945
Y-1	Jasmine Yellow	2946
Y-4	Golden Fawn	2635
Y-6	Inca Gold Poly	2738
Y-7	Taxi Yellow	81746
Y-9	Spanish Gold Poly	2739
999	Special Order Colors	

TWO-TONES: First two digits are accent or roof colors, second two digits are basic body color.

1978 CHRYSLER, CORDOBA & LE BARON

EW1	Spinnaker White	2033
EY7	Taxi Yellow	81746
KY4	Golden Fawn	2635

LY9	Spanish Gold Poly	2739
MU3	Caramel Tan Poly	2856
PB3	Cadet Blue Poly	2935
PB9	Starlight Blue Sunfire Poly	2938
PY1	Jasmine Yellow	2946
RA1	Dove Gray	3015
RA9	Charcoal Gray Sunfire Poly	3017
RF3	Mint Green Poly	3019
RA3	Wedgewood Grey Poly	33287
RF9	Aug. Green Sunfire Poly	3018
RR7	Tap. Red Sunfire Poly	3020
RT9	Sable Sunfire Poly	3014
RY3	Classic Cream	3021
TX9	Black	9300,9000

1979 CHRYSLER

EW1	Spinnaker White	2033(C,D,P)
EW7	Taxi Yellow	81746
MV1	Spitfire Orange	2858(C,D)
PB3	Cadet Blue Poly	2935(C,D,P)
RA1	Dove Gray	3015(C,D,P)
RA2	Pewter Gray Poly	3016(C,D,P)
RT9	Sable Tan Sunfire Poly	3014(C,D,P)
SA5	Smoked Gray Poly (Two-tone)	33264(H,O)
SA6	Oxford Gray(Two-tone)	33285(C,D)
SB7	Ensign Blue Poly	3144(C,D,P)
SC2	Frost Blue Poly	3145(C,D)
SC9	Nightwatch Blue	3146(C,D)
SG4	Teal Frost Poly	3147(C,D,P)
SG8	Teal Green Sunfire Poly	3148(C,D,P)
SL1	Designers Cream (Two-tone)	24569(C,D)
SL2	Designers Beige (Two-tone)	24570(C,D)
SQ6	Turquoise Poly	3152(H,O)
SR5	Chianti Red	3149(C,D,P)
SR8	Regent RedSunfire Poly	3150(C,D,P)
SR9	Garnet Red Sunfire Poly	3151(C,D,H,O)
SS1	Pearl Gray(Two-tone)	90107(C,P)
ST1	Light Cashmere	3142(C,D,P)
ST5	Medium Cashmere Poly	3143(C,D,P)
SV3	Flame Orange	3153(H,O)
SY1	Linen Cream (Two-tone)	3358(C,P)
SY2	Light Yellow	82388(D,P)
SY4	Bright Yellow	82398(H,O)
TX9	Black	9300(C,D,P)

C-Chrysler, P-Plymouth, D-Dodge, H-Horizon, O-Omni

1980 CHRYSLER, DODGE, PLYMOUTH

EW1	Spinnaker White	2033
SC2	Frost Blue Poly	3145
SC9	Nightwatch Blue	3146
SG4	Teal Frost Poly	3147
SL1	Designer's Cream	24569
SL2	Designer's Beige	24570
ST1	Light Cashmere	3142
SY4	Bright Yellow	3274(D,P)
TA3	Burnished Silver Poly	3261
TB6	Graphic Blue	3262(D,P)
TD2	Light Heather Gray	3263
TD3	Light Heather Gray Poly	3264
TD6	Dark Heather Brown	51053
TG6	Teal Tropic Green Poly	3265
TM7	Crimson Red Poly	3266
TM9	Baron Red	3267
TR4	Graphic Red	3268
TT4	Natural Suede Tan	3269
TT7	Black Walnut Poly	3273
TT8	Mocha Brown Poly	3270
TW2	Bravo White	90131(C,D)
TX9	Black	9000,9300

C-Chrysler, P-Plymouth, D-Dodge

1981 CHRYSLER, DODGE, PLYMOUTH

SC9	Nightwatch Blue	3146
SL1	Designer's Cream	24569
SY1	Light Cashmere	3142
SY1	Linen Cream	3358
TA3	Burnished Silver Poly	3261
TD2	Light Heather Gray	3263
TM9	Baron Red	3267
TR4	Graphic Red	3268
TT4	Natural Sude Tan	3269
TT6	Ginger	3271
VA2	Sterling Silver Poly	33444(C,P)
VB7	Vivid Blue Poly	3344
VC3	Day Star Blue Poly	3345
VD1	Driftwood Gray	33414
VD4	Heather Mist Poly	3346
VF2	Lt. Seaspray Green Poly	3347
VF8	Glencoe Green Poly	3348
VH5	Auburn Mist Poiy	33445(C,P)
VH9	Mahogany Poly	3301
VK8	Spice Tan Poly	3349
VM8	Morocco Red	72407
VT3	Manila Cream	24838
VT5	Light Caramel Tan	3351
VT9	Coffee Brown Poly	24906

VW3	Pearl White	3352
VY2	Sunlight Yellow	3353
VY5	Graphic Yellow	3354
DX9	Low Gloss Black	9440
TX9	Formal Black	9300

C-Chrysler, P-Plymouth, D-Dodge

1982 CHRYSLER, DODGE, PLYMOUTH

AA9	Charcoal Poly	3495
AB5	Ensign Blue Poly	9496(D,P)
AC8	Dark Blue Poly	3497(D,P)
AM6	Medium Red Poly	3498(C)
AT7	Medium Tan Poly	3499(C)
SC9	Nightwatch Blue	3146(C,D)
TA3	Burnished Silver Poly	3261
TR4	Graphic Red	3268(D,P)
TT4	Suede Tan	3269(D,P)
VA1	Sterling Silver Poly	33468(C)
VC1	Daystar Blue	15472(C)
VC3	Daystar Blue Poly	3345
VC4	Medium Blue Poly	3600(C)
VD4	Heathermist Poly	3346(C)
VF2	Light Seaspray Green Poly	3347(D,P)
VF7	Medium Seaspray Poly	45556(D,P)
VH2	Light Auburn Poly	33516(C)
VH9	Dark Mahogany Poly	3301(C,D)
VK8	Spice Tan Poly	3349
VM8	Medium Crimson Red	72407
VT3	Manila Cream	3518
VT5	Medium Tan	3351(C,D)
VW3	Snow White	3352
DX9	Flat Black	9440(D,P)
TX9	Black	9300

C-Chrysler, P-Plymouth, D-Dodge

1983 CHRYSLER, DODGE, PLYMOUTH

AA9	Charcoal Gray Poly	33645
BA2	Silver Crystal Poly	3495
BB6	Silver Crystal Coat	3557
BB6	Santa Fe Blue Crystal Coat	3619
BC2	Light Blue Poly	15623
BK6	Spice Poly	25073
BL2	Beige Poly	25174
BL4	Beige Crystal Coat	3558
BL5	Beige Sand	25074
BM5	Crimson	3559
BT8	Sable Brown	3560
CA1	Silver Crystal Coat	3556
DX9	Flat Black	9440
SC9	Nightwatch Blue	3146
SV3	Impact Orange	3153

TA3	Burnished Silver Poly	3261
TR4	Graphic Red	3268
TX9	Black	9300
VC4	Glacier Blue Crystal Coat	3600
VW3	Pearl White"	3352

SHELBY SPECIAL EDITION - TWO TONE:

BB6	Santa Fe Blue Crystal Coat	3619
CA1	Silver Crystal Coat	3556

1984 CHRYSLER, DODGE, PLYMOUTH

AA9	Charcoal Gray Poly	3495
AC8	Navy Blue Poly	3497
BB6	Santa Fe Blue Crystal Coat	3619
BC2	Light Blue Poly	15623
BK6	Spice Poly	25073
BL2	Beige Poly	25174
BL4	Beige Crystal Coat	3558
BL5	Beige Sand	25074
BM5	Crimson	3559
BT8	Sable Brown	3560
CA1	Radiant Silver Crystal Coat	3556
CA6	Charcoal Pearl Coat	3635
CC6	Gunmetal Blue Pearl Coat	3631
CK5	Saddle Brown Crystal Coat	25200
CT6	Mink Brown Pearl Coat	3634
RR7	Canyon Red Sunfire Poly	3020
SC9	Nightwatch Blue	3146
TA3	Silver Poly	3261
TR4	Graphic Red	3268
VC4	Glacier Blue Crystal Coat	3600
VW3	Pearl White	3352
TX9	Black	9300

1985 CHRYSLER DODGE, PLYMOUTH

BA5	Charcoal Pearl Coat	3702
BB6	Santa Fe Blue Poly	3619
CA1	Radiant Silver Poly	3556
CC6	Gunmetal Blue Pearl Coat	3631
CR3	Carrera Red	3703
CR6	Garnet Pearl Coat	3633
CT6	Mink Brown Pearl Coat	3634
DB1	Ice Blue Poly	3704
DB9	Nightwatch Blue	3705
DE5	Desert Bronze Pearl Coat	3706
DK7	Spice Poly	3707
DM6	Crimson Red	3708
DR5	Graphic Red	3709
DT3	Cream	3710
DT4	Gold Dust	3714

DW2	White	3712
DX8	Black	9300
VC4	Glacier Blue Poly	3600

1986 CHRYSLER, DODGE, PLYMOUTH

CA1	Radiant Silver Poly	3556
CC6	Gunmetal Blue Pearl Coat	3631
CR3	Flash Red	3703
CR6	Garnet Pearl Coat	3633
CT6	Mink Brown Pearl Coat	3634
DB1	Ice Blue Poly	3704
DB9	Nightwatch Blue	3705
DM6	Crimson Red	3708
DR5	Graphic Red	3709
DT4	Gold Dust Poly	3714
DW2	White	3712
DX8	Black	9700
EC7	Twilight Blue Pearl Coat	3831
EE2	Misty Rose Pearl Coat	3832
EE9	Dk.Cordovan PearlCoat	3833
ET1	Lt.Cream	3834
ET5	Golden Bronze Pearl Coat	3835
FS8	Charcoal Pearl Coat	3836

1987 CHRYSLER, DODGE, PLYMOUTH

BA5	Charcaal Gray Pearl Coat	3702
CA1	Radiant Silver Poly	3556
CC6	Gunmetal Blue Pearl Coat	3631
CR3	Flash Red	3703
CR6	Garnet Pearl Coat	3633
DB1	Ice Blue Poly	3704
DB9	Nightwatch Blue	3705
DM6	Crimson Red	3708
DR5	Graphic Red	3709
DT4	Gold Dust Poly	3714
DW2	White	3712
DX8	Black	9700
EC7	Twilight Blue Pearl Coat	3831
EE2	Misty Rose Pearl Coat	3832
EE9	Dk.Cordovan PearlCoat	3833
ET1	Lt.Cream	3834
ET5	Golden Bronze Pearl Coat	3835
FS8	Charcoal Pearl Coat	3836
FU7	Chestnut Brown Pearl Coat	3933

1988 CHRYSLER, DODGE, PLYMOUTH

CA1	Radiant Silver Poly	3556
CR3	Flash Red	3703
DB1	Ice Blue Poly	3704
DR5	Graphic Red	3709
DX8	Black	9700

EC7	Twilight Blue Pearl Coat	3831
EE2	Lt. Rosewood Pearl Coat	3832
EE9	Dk. Cordovan Pearl Coat	3833
ET1	Lt. Cream	3834
ET5	Golden Bronze Pearl Coat	3835
FA7	Charcoal Poly	4053
FM9	Black Cherry Pearlcoat	4043
FS8	Charcoal Pearl Coat	3836
FW9	White	3700
FX6	Black	9700
GB4	Daytona Blue Poly	4044
GG7	Dark Forest Green Poly	4045
GK4	Med. Suede Poly	4046
GK9	Dark Suede Poly	4047
GL2	Light Pewter Pearl Coat	4048
GL8	Dark Pewter Pearl Coat	4051
GM4	Claret Red Pearl Coat	4052

1989 CHRYSLER, DODGE, PLYMOUTH, EAGLE

CA1	Radiant Silver Poly	3556
CR3	Flash Red	3703
DB1	Ice Blue Poly	3704
DX8	Black	9700
DR5	Graphic Red	3709
DX8	Black	9700
EC7	Twilight Blue Pearl Coat	3831
EE2	Lt. Rosewood Pearl Coat	3832
EE9	Dk. Cordovan Pearl Coat	3833
FA7	Charcoal Gray Poly	4053
FM9	Black Cherry Pearlcoat	4043
FR9	Garnet Red Poly	4120
FS8	Charcoal Pearl Coat	3836
FX6	Classic Black	9700(E)
GB4	Daytona Blue Poly	4044
GK4	Med. Suede Poly	4046
GK9	Dark Suede Poly	4047
GL2	Light Pewter Pearl Coat	4048
GL3	Light Taupe Poly	4049
GL6	Medium Taupe Poly	4050
GL8	Dark Pewter Pearl Coat	4051
GM4	Claret Red Pearl Coat	4052
GW7	Bright White	4037
HA2	Sterling Silver Poly	3594(E)
HC1	Diamond Blue Poly	4142
HB8	Midnight Blue Poly	4141
HC1	Diamond Blue Poly	4142
HD1	Platinum Silver Poly	4143
HD8	Dark Quartz Gray Poly	4144
HF5	Dark Driftwood Poly	4035
HJ3	Sunlight Yellow	4145
HM2	Dark Cherry	4038

HM7	Dusty Rose	4147
HQ1	Platinum Blue Poly	4151
HQ7	Aquamarine Blue Poly	4149
HQ9	Dark Baltic Blue Poly	4031
HR1	Exotic Red	4148
HS3	Dove Gray Poly	4029
HV3	Lt. Champagne Poly	4150
HWB	Pearl White	4027(E)
HY3	Sand Poly	4138
	E-EAGLE	

1990 CHRYSLER CORPORATION

CA1	Radiant Silver Poly	3556
CR3	Flash Red	3703
DB1	Ice Blue Poly	3704
DX8	Black	9700
EC7	Twilight Blue	3831
EE2	Light Rosewood	3832
FA7	Charcoal Gray Poly	4053
FM9	Black Cherry Pearlcoat	4043
FR9	Garnet Red Poly	4120
FS8	Charcoal Pearl Coat	3836
GB4	Daytona Blue Poly	4044
GK4	Med. Suede Poly	4046
GK9	Dark Suede Poly	4047
GL3	Light Taupe Poly	4049
GL6	Medium Taupe Poly	4050
GM4	Claret Red Pearl Coat	4052
GW7	Bright White	4037
HA2	Sterling Silver Poly	3594
HB8	Midnight Blue Poly	4141
HC1	Diamond Blue Poly	4142
HD1	Platinum Silver Poly	4143
HD8	Dk. Quartz Gray Poly	4144
HE4	Colorado Red	3932
HM2	Dk. Cherry	4038
HM7	Dusty Rose	4038
HQ1	Platinum Blue Poly	4151
HQ7	Aquamarine Blue Poly	4149
HQ9	Dark Baltic Blue Poly	4031
HR1	Exotic Red	4148
HS3	Dove Gray Poly	4029
HV3	Lt. Champagne Poly	4150
HY3	Sand Poly	4138
JB2	Lt. Spectrum Blue Poly	4192
JB5	Med. Blue Gray	4193
JH6	Lt. Mahogany Poly	4194
JH8	Dk. Mahogany Poly	4195
JT5	Medium Tundra Poly	4196

1991 CHRYSLER CORPORATION

B32	Lt. Spectrum Blue Poly	4192(L)
B35	Banzai Blue Poly	4271(P,D,L)
DX8	Black	9700(C,P,D)
EC7	Twilight Blue	3831(P,D)
EE2	Light Rosewood	3832(D)
FM9	Black Cherry Pearlcoat	4043(C,P,D)
FR9	Garnet Red Poly	4120
FS8	Charcoal Pearl Coat	3836(C)
GM4	Claret Red	4052(C,P,D)
GW7	Bright White	4037(C,P,D)
HA2	Sterling Silver Poly	3594(D)
HB8	Midnight Blue Poly	4141(C,D)
HC1	Diamond Blue Poly	4142(C,P,D)
HD1	Platinum Silver Poly	4143(C,P,D)
HD8	Dk. Quartz Gray Poly	4144(C,P,D)
HE4	Colorado Red	3932(D)
HM7	Dusty Rose	4038(C,P,D)
HM3	Raspberry	4190(P,D)
HM7	Dusty Rose	4147(C,P,D)
H16	Med. Quartz Poly	4321(L)
H18	Lt. Mynx	4189(L)
JB2	Lt. Spectrum Blue Poly	4192(P,D)
JB5	Med. Blue Gray	4193(C,D)
JH6	Lt. Mahogany Poly	4194(C)
JH8	Dk. Mahogany Poly	4195(C)
KBD	Light Blue Satin Glow	4269(C,P,D)
KBF	Med. Water Blue Satin Glow	4270(C,P,D)
KC3	Banzai Blue Poly	4271(P,D)
KQ2	Glamour Turquoise Poly	4272(P,D)
KRB	Radiant Fire Red	4573(C,P,D)
KS7	Dark Silver Poly	4274(C,D)
KTA	Safari Brown Satin Glow	4275(P,D)
K13	Tennessee Blue Poly	4204(L)
R11	Flash Red	4186(L)
W12	Bright White	4037(L)
X13	Black	9700(L)

C-Chrysler, P-Plymouth, D-Dodge, L-Laser

1992 CHRYSLER CORPORATION

DX8	Black	9700(C,D,P, DC,PV)
FM9	Black Cherry Pearlcoat	4043(C,P,D,)
GM4	Claret Red	4052(C,P,D)
GW7	Bright White	4037(C,P,D,)
HA2	Sterling Silver Poly	3594(D)
HB8	Midnight Blue Poly	4141(C,D)
HC1	Diamond Blue Poly	4142(C,D)
HD8	Dk. Quartz Gray Poly	4144(C,D,)
HM3	Raspberry	4190(P,D)

JH6	Lt. Mahogany Poly	4194(C)
JH8	Dk. Mahogany Poly	4195(C)
KBD	Light Blue Satin Glow	4269(P,D)
KBF	Med. Water Blue Satin Glow	4270(C,P,D)
KC3	Banzai Blue Poly	4271(P,D)
KRB	Radiant Fire Red	4274(C,P,D)
KS7	Dark Silver Poly	4274(D)
KV4	Light Champagne Poly	4320(C,P,D)
LDA	Bright Silver Quartz Poly	4443(C,D,P)
LG6	Beryl Green Pearlcoat	4444(C,D)
LP5	Teal Pearlcoat	4445((D,P)
LQE	Aqua Pearlcoat	4446(C,D,P)
LRF	Radiant Red	4447(D)
LRN	Viper Red	73840(D)

C-Chrysler, D-Dodge, P-Plymouth

1993 CHRYSLER CORPORATION

PBD/KBD	Light Blue Satin Glow	4269(C,D,P)
PBF/KBF	Med. Water Blue Satin Glow	4270(C,D,E,P)
PB8/HB8	Midnight Blue Poly	4141(C,D)
PC1/HC1	Diamond Blue Poly	4242(C,D)
PC3/KC3	Banzai Blue Poly	4271(D,P)
PDA/LDA	Bright Silver Quartz Poly	4443(C,D)
PFA/MFA	Lt. Driftwood Satin Glow	4569(C,D,E,P)
PGF	Emerald Green Pearlcoat	4639(C,D,E,P)
PGQ/MGQ	Emerald Green Pearlcoat	47260 (Viper only)
PG6/LG6	Beryl Green Pearlcoat	4444(C,D)
PH6/JHc	Light Mahogany Poly	4194(C)
PJE/MJE	Dandelion Yellow	83431(Viper Only)
PJ8/MJ8	Char-Gold Satin Glow	4677(C,D,E)
PMB/MMB	Wildberry Pearlcoat	4678(C,D,P,)
PM3/HM3	Rasberry Pearlcoat	4190(D,P)
PM4/GM4	Claret Red Pearlcoat	4052(C,D)
PP5/LP5	Teal Pearlcoat	4445(C,D,E,P)
PQE/LQE	Aqua Pearlcoat	4446(C,D,P)
PRB/KRB	Radiant Fire	4273(C,D,E,P)
PRF/LRF	Radiant Red	4447(C,D,E,P)
PRN/LRN	Viper Red	73840(Viper Only)
PV4/KV4	Light Champagne Poly	4320(C,D)
PW7/GW7	Bright White	4037(C,D,E,P,)
PX3/MX3	Black	9874(Viper Only)
PX8/DX8	Black	9700(C,D,E,P)

C-Chrysler, D-Dodge, P-Plymouth, E-Eagle

1994 CHRYSLER CORPORATION

PCH	Brilliant Blue Pearlcoat	4784(D,P)
PC3/KC3	Banzai Blue Poly	4271(D,P)
PC5	Light Iris Pearlcoat	4788(D,P)
PFA/MFA	Lt. Driftwood Satin Glow	4569(C,D,P)
PGF	Emerald Green Pearlcoat	4639(C,D,E,P)
PGQ/MGQ	Emerald Green Pearlcoat	47260 (Viper only)
PGF	Emerald Green Pearlcoat	4785(D,P)
PJE/MJE	Dandelion Yellow	83431(Viper
PJ8/MJ8	Char-Gold Satin Glow	4677(C,D,E)
PMB/MMB	Wildberry Pearlcoat	4678(C,D,P)
PME	Wildberry Pearlcoat	4863(C,D,P)
PM3/HM3	Rasberry Pearlcoat	4190(D,P)
PM9/FM9	Black Cherry Pearlcoat	4043(C,D,E)
PPD	Teal Pearlcoat	4864(C,D,P)
PP2	Pale Blue Clearcoat	4787(D,P)
PP5/LP5	Teal Pearlcoat	4445(C,D,E,P)
PQE/LQE	Aqua Pearlcoat	4446(D,P)
PQK	Aqua Pearlcoat	4786(D,P)
PQ2/KQ2	Glamour Turquoise Poly	4272(D,P)
PRB/KRB	Radiant Fire Clearcoat	4273(C,D,E,P)
PRE	Strawberry Pearlcoat	4491(D,P)
PRF/LRF	Radiant Red Poly	4447(C,D,E,P)
PRN/LRN	Viper Red Clearcoat	73840 (Viper Only)
PR4	Flame Red Clearcoat	4679(D,P)
PS4/MS4	Bright Platinum Poly	4820(C,D,E)
PX3/MX3	Black Clearcoat	9874(Viper Only)
PX8/DX8	Black Clearcoat	9700(C,D,P)

C-Chrysler, D-Dodge, P-Plymouth, E-Eagle

1995 CHRYSLER CORPORATION

DX8	Black Clearcoat	9700(C,D,P,E)
GW7	Bright White Clearcoat	4037(C,D,P,E)
KRB	Radiant Fire Clearcoat	4273(C)
LRN	Viper Red Clearcoat	73840 (Viper Only)
MFA	Light Driftwood Satin Glow	4569(C,D,P,E)
MGQ	Viper Emerald Green	47260 (Viper only)
MJE	Viper Bright Yellow	83431 (Viper Only)
MJ8	Char-Gold Satin Glow	4677(C,D,E)
MMB	Wildberry Pearlcoat	4678(C,D,P,E)
MS4	Brite Platinum Silver Poly	4820(C,D,E,P)
MX3	Viper Black	9874(Viper Only)
PCH	Brilliant Blue Pearlcoat	4784(N)
PC5	Light Iris Pearlcoat	4788(D,P,N)
PGS	Emerald Green Pearlcoat	4785(C,D,P,N)
PME	Wildberry Pearlcoat	4863(C,D,P)
PQK	Aqua Pearlcoat	4786(N)
PRE	Strawberry Pearlcoat	4791(N)
PR4	Flame Red Clearcoat	4679(N)
RB3	Medium Blue Pearlcoat	4962(C,D,E,P)
RBK	Medium Blue Pearlcoat	4963(D,P)
RC4	Lapis Blue Clearcoat	4935(N)
REF	Light Rosewood Satin Glow	4965(D,P)
REG	Dark Rosewood Pearlcoat	4966(D,P)
RF2	Nitro Yellow-Green	4940(N)
RJM	Light Silverfern Pearlcoat	4968(D,P)
RJP	Medium Fern Pearlcoat	4969(D,P)
RMH	Wild Orchid Pearlcoat	4970(C,D,P,E)
RMK	Wild Orchid Pearlcoat	4971(D,P)
RP6	Island Teal Satin Glow	4972(C,D,P,E)
RPE	Spruce Pearlcoat	4973(C,D,P,E)
RRC	Metallic Red Pearlcoat	4974(D,P)

C-Chrysler, D-Dodge, P-Plymouth, E-Eagle

Chapter 6

DODGE 1941-1978

PAINT CODE	COLOR	PPG CODE
1941 DODGE		
103	Black	9000
207	Regimental Blue	10027
209 & 210	Seaplane Blue No. 3	10024
307	Harbor Green Poly	40048
309	Fairway Green	40010
	(Original Color)	
	(To match weathered car)	40081
323	Provencal Green	40011
408	Gold Beige Poly	20033
505 & 506	Speedwing Gray	30007
519 Body	Pursuit Gray	20010
Wheels	Reconnaissance Red	70044
605	Regal Maroon	50037
607	Flare Red	50008
807	Dodge Gunmetal	30025
	(Original Color)	
	(To match weathered car)	30054
809	Tennis Cream	80040
	(Original Color)	
	(To match weathered car)	80039

TWO-TONES

906	U 40013	908 & 923	U 10023	912	U 30064
	L 40010		L 10024		L 40048
907	U 30007	909 & 924	U 10024		
	L 50037		L 30064		

For upper color on two-tone #912 use 30178 for serial number higher than 30,438 or 150.

1942 DODGE

103 & 104	Black	9000
205 & 207	Patrol Blue	10001
209 & 211	LaPlata Blue	10024
213	Bombardier Blue	10010
305 & 306	Windward Green	40018
308 & 309	Oricoco Green	40020
311 & 312	Forest Green	40005
403 & 404	Tampico Beige	20009
503 & 505	Fortress Gray	30002
605 & 606	Military Maroon	50037
608	Squad Red	70004
805 & 806	Cadet Gray	30006
808	Panama Sand	80001

TWO-TONES

908 & 909	U 10001			916 & 917	U 40005
	L 10024				L 40018
910 & 913	U 10024			918 & 919	U 40018
	L 10001				L 40005
912 & 913	U 40020			920 & 921	U 30006
	L 40018				L 30002
914 & 915	U 40018			922 & 923	U 30002
	L 40020				L 30006

1946 - 1947 - 1948 DODGE

103 & 104	Black	9000
205 - 207	Patrol Blue	10001
209 - 211	LaPlata Blue	10024
305 - 306	Windward Green	40018
308 & 309	Orinoco Green	40020
311 & 312	Forest Green	40005
324	Gypsy Green Sympho	40559
403, 404 & 412	Stone Beige No. 1	20006
505 & 506	Opal Gray No. 4	30139
508 - 510	Fortress Gray	30002
605 & 606	Military Maroon	50037
608	Air Cruiser Red	50009
610	Squad Red	70004
803	Panama Sand	80001

1949 DODGE

501	Black	9000
505	LaPlata Blue	10024
506	Tunis Blue Sympho	10346
520	Island Green	40633
521	Hunter Green	40805
522	Gypsy Green Sympho	40559
535	Granite Gray	30575
536	French Gray	30572
545	Stone Beige	20006
546	Cairo Tan Poly	20520
560	Monarch Maroon	50137
561	Cadet Red	70200
562	Air Cruiser Red	50009
565	Victoria Ivory	80405

1950 DODGE

506	Dominion Blue	10022
545	Nassau Beige	20555
546	Burma Tan Poly	20680

All other 1950 colors were carried over from 1949 production. See 1949 listing for the following:

501 - 9000, 505 - 10024, 520 - 40633, 521 - 40805, 522 - 40559, 535 - 30575, 536 - 30572, 560 - 50137, 561 - 70200, 562 - 50009, 565 - 80405.

1951 DODGE

501	Black	9000
505	Pitcairn Blue Lt.	10601
506	Dominion Blue	10022
520	Sea Mist Green	41007
521	Ceram Green Med	41075
522	Gypsy Green Sympho	41076
523	Manchu Green Poly	41095
524	Silhouette Green	40861
525	Manchu Green Poly (Estate Wagon)	41095
535	Heron Gray	30589
536	Dover Gray	30799
545	Nassau Beige	20555
546	Kachina Bronze Lt. Poly	20747
547	Fawn Beige	20773
548	Oakwood Bronze Poly	20405
560	Monarch Maroon	50137
561	Troubador Red	50215
562	Aircruiser Red	50009
565	Victoria Ivory	80405
566	Jungle Lime	40929

TWO-TONES

570	U 30799	573	U 40789	578	U 41095
	L 10601		L 41076		L 40861
571	U 20747	574	U 20535		
	L 20555		L 50137		
572	U 30592	575	U 41066		
	L 30799		L 40929		

1952 DODGE

501	Black	9000
505	Pitcarin Blue	10601
507	Fairfax Blue Poly	10822
520	Seamist Green	41007
524	Silhouette Green (Shade Similar to 41294)	40861
522	Gypsy Green Sympho	41076
535	Heron Gray	30589
536	Dover Gray	30799
545	Nassau Beige	20555
548	Oakwood Bronze Poly (Shade Similar to 20871)	20405
562	Aircruiser Red	50009
564	Fiesta Maroon Poly	50284
565	Victoria Ivory	80405

1953 DODGE

501	Black	9000
505	Bimini Blue	10748
507	Fairfax Blue Poly	10822
520	Seamist Green	41007
524	Silhouette Green	41294
522	Gypsy Green Sympho	41076
535	Heron Gray	30589
536	Dover Gray	30799
545	Nassau Beige	20905
548	Oakwood Bronze Poly	20871
562	Esquire Red	70408
564	Fiesta Maroon Poly	50284
565	Shoreham Ivory	80597

TWO-TONES

580	U 10822	582	U 41376	584	U 31009
	L 10987		L 41377		L 10144
581	U 41377	583	U 41240		
	L 41376		L 41173		

1954 DODGE

501	Jewell Black	9000
505	Bermuda Blue	11076
506	Bedford Blue	11081
507	Lancaster Blue Poly	11072
515	Willow Green	41363
516	Berkshire Green	41540
517	Cumberland Green Poly	41179
530	Dawn Gray	31094
531	Wing Gray	31126
540	Sunsand	80625
550	Esquire Red	70408
556	Pace Car Yellow	80690

TWO-TONES

560	U 31094	565	U 11081	570	U 41179
	L 11072		L 80625		L 80625
561	U 80625	566	U 41363	571	U 80625
	L 11081		L 41179		L 70408
562	U 10081	567	U 41363	572	U 70408
	L 11076		L 41540		L 80625
563	U 31094	568	U 80625	573	U 70408
	L 31126		L 41540		L 31094
564	U 11081	569	U 41179		
	L 31094		L 41363		

These obsolete 1953 colors were used on 1954 models until stock was exhausted.

508	Bimini Blue	10748
520	Seamist Green	41007
535	Heron Gray	30589

| 536 | Dover Gray | 30799 |
| 555 | Shoreham Ivory | 80597 |

TWO-TONES

574	U 80597	580	U 80697	583	U 41584
	L 41007		L 60159		L 80697
577	U 80597	581	U 60159	585	U 9000
	L 70408		L 80697		L 80597
579	U 70408	582	U 80697	586	U 80597
	L 80597		L 41584		L 10748

1955 DODGE

501	Jewell Black	9000
505	Halo Blue	11246
506	Parisian Blue	11297
507	Admiral Blue Poly	11248
515	Chiffon Green	41653
516	Emerald Green	41654
517	Satin Green Poly	41655
530	Cashmere Gray	31252
550	Heather Rose	70522
551	Cameo Red	70523
552	Regal Burgundy Poly	50399
555	Sapphire White	21134
556	Fantasy Yellow	80729

TWO-TONES

560	U 11246	566	U 41654	572	U 21134
	L 11248		L 41653		L 9000
561	U 11248	567	U 70523	573	U 21134
	L 11246		L 31252		L 80729
562	U 21134	568	U 9000	574	U 9000
	L 11297		L 70523		L 80729
563	U 41654	569	U 9000	575	U 21134
	L 41655		L 70522		L 50399
564	U 41655	570	U 70523		
	L 41654		L 21134		
565	U 41655	571	U 70522		
	L 41653		L 21134		

THREE-TONES

560-2	U 11246	566-2	U 41654	572-4	U 21134
	L 11248		L 41653		L 9000
	21134		21134		70523
561-2	U 11248	567-3	U 70523	573-3	U 21134
	L 11248		L 31252		L 80729
	21134		9000		9000
562-3	U 21134	568-2	U 9000	574-2	U 9000
	L 11297		L 70523		L 80729
	9000		21134		21134
563-2	U 41654	569-2	U 9000	575-3	U 21134
	L 41655		L 70522		L 50399
	21134		21134		9000
564-2	U 41655	570-3	U 70523		
	L 41654		L 21134		
	21134		9000		
565-2	U 41655	571-3	U 70522		
	L 41655		L 21134		
	21134		9000		

Cars painted in three-tone combinations can be identified by the suffix (2,3 or 4) shown after the paint code number. The suffix number (2,3 or 4) will identify the third color as follows.

2	Sapphire White, 21134
3	Jewell Black, 9000
4	Cameo Red, 70523

Example: 560-2 is a three-tone combination the figure 560 identifying HALO BLUE 11246 and ADMIRAL BLUE POLY. 11248 as the two-tone colors and the suffix 2 identifying SAPPHIRE WHITE 21134 as the third tone.

1956 DODGE

501	Jewell Black	9000
505	Wedgewood Blue	11442
506	Royal Blue Poly	11371
515	Aquamarine	41871
516	Neptune Green Poly	41870
517	Sea Foam Green	41850
518	Jade Green Poly	41793
540	Cloud Gray	31225
541	Iridescent Charcoal Poly	31325
550	Chinese Rose	70635
551	Oriental Coral	70648
552	Garnet Poly	50427
555	Sapphire White	21134
556	Crown Yellow	80795

Two-Tones Only:

	Gallant Gold Poly	21293
	Misty Orchid	50450
	Regal Orchid	50449

TWO-TONES

560	U 21134	566	U 41850	572	U 21134
	L 9000		L 41793		L 70648
560-1 Same as 560					
566-1 Same as 566					
572-1 Same as 572					
560-4	R 70648	566-3	R 9000	572-3	R 9000
	S 21134		S 41850		S 21134
	L 9000		L 41793		L 70648
561-1	U 11371	567	U 31325	573	U 9000

```
        L 11442          L 31225            L 50427
561-1 Same as 561
567-1 Same as 567
573-1 Same as 573
561-2 R 21134    567-2 R 21134    573-2 R 21134
      S 11371          S 31325          S 9000
      L 11442          L 31225          L 50427
562  U 21134     568  U 31325     574  U 31325
      L 11371          L 70635          L 80795
562-1 Same as 562
568-1 Same as 568
574-1 Same as 574
562-3 R 9000     568-2 R 21134    574-2 R 21134
      S 21134          S 31325          S 31325
      L 11371          L 70635          L 80795
563  U 9000      569  U 50427     575  U 21134
      L 41871          L 70635          L 80795
563-1 Same as 563
569-1 Same as 569
575-1 Same as 575
563-2 R 21134    569-3 R 9000     575-3 R 9000
      S 9000           S 50427          S 21134
      L 41871          L 70635          L 80795
564  U 21134     570  U 21134
      L 41870          L 70635          R 21134
564-1 Same as 564
570-1 Same as 570                       S 21293
564-3 R 9000     570-3 R 9000     L 21134
      S 21134          S 21134
      L 41870          L 70635     U 50450
565  U 41793     571  U 31325     L 50449
      L 41850          L 70648
565-1 Same as 565
571-1 Same as 571
565-2 R 21134    571-2 R 21134
      S 41793          S 31325
      L 41850          L 70648
```

NOTE: The one digit figure after the combination code indicates the following:
1 - Deluxe Model
2 - Sapphire White Used as the Roof Color
3 - Jewell Black Used as the Roof Color
4 - Oriental Coral Used as Roof Color

1957 DODGE

AAA	Jewel Black	9000
BBB	Ice Blue	11531
CCC	Velvet Blue Poly	11529
DDD	Misty Green	41960
EEE	Forest Green Poly	41496
FFF	Moonstone Gray	31410

GGG	Metallic Charcoal Poly	31389
HHH	Flame Red	70691
KKK	Sunshine Yellow	80850
LLL	Glacier White	8131
For cars built on West Coast us Cloud White 8036		
MMM	Turquoise	11532
NNN	Tropical Coral	70698
For cars built on West Coast use Gauguin Red 70693		
PPP	Gallant Gold Poly	21293
SSS	Heather Green	41961

Color combinations can be identified by the arrangement of the paint code letters. For example -

SINGLE TONE "T" INDICATES STANDARD TWO-TONES AAA,BBB, Etc. TLC
"L"=Roof, Glacier White, 8131
"C"=Lower, Velvet Blue Poly, 11529

SPECIAL TWO-TONES LCC
"L"=Roof and Fins, Glacier White, 8131
"CC"=Saddle and Lower, Velvet Blue Poly, 11529

DELUXE TWO-TONES CLC
First Letter "C"=Roof and Fins, Velvet Blue Poly, 11529
Second Letter "L"=Saddle, Glacier White, 8131
Third Letter "C"=Lower, Velvet Blue Poly, 11529

1958 DODGE

AAA	Ebony	9000
BBB	Wedgewood	11684
CCC	Sapphire Poly	11596
DDD	Navy	11679
EEE	Mint	42176
FFF	Moss Poly	42128
KKK	Silver Poly	31539
LLL	Charcoal Poly	31537
MMM	Beige	20952
NNN	Sand Poly	21562
PPP	Crimson	70791
SSS	Copper Poly	21569
UUU	Sunshine	80062
XXX	Eggshell	8131
ZZZ	Light Gold Poly	21519

On two-tone and insert combinations, the first letter indicates roof and fins, second letter saddle and the third letter indicates lower color.

1959 DODGE

AA-1	Jet Black	9000
BB-1	Blue Diamond	11813
CC-1	Star Sapphire Poly	11812
EE-1	Aquamarine	42315
FF-1	Jade Poly	42295
JJ-1	Turquoise	11720

LL-1		Silver Poly	31539
MM-1		Pewter Poly	31660
NN-1		Rose Quartz	70884
PP-1		Coral	70885
RR-1		Ruby	70791
UU-1		Biscuit	21642
WW-1		Mocha Poly	21587
XX-1		Pearl	8131
YY-1		Canary Diamond	81057
*JJJ		Frosted Turquoise Poly	11923
*RRR		Poppy	70911
*TTT		Paris Rose Poly	70923

* 1958 Spring Color

Color combinations can be identified by the arrangement of the paint code letters followed by a paint code number.

For example: CC-1 Single Tone

CB-2 Fin Only

First Letter is Fin Color

Second Letter Balance of Car

CB-3 Fin & Lower

First Letter Fin & Lower

Second Letter Balance of Car

CB-4 Roof, Fins & Lower

First Letter is Roof, Fins & Lower

Second Letter Balance of Car

1960 DODGE

AA-1		Raw Sienna	31125
BB-1		Raven	9000
CC-1		Azure	11988
DD-1		Mediterranean Poly	12044
FF-1		Spray	42539
GG-1		Spruce Poly	42538
HH-1		Cactus Poly	42540*
JJ-1		Frost Turquoise	11787**
KK-1		Teal Poly	42263
LL-1		Cloud	30125
MM-1		Pewter Poly	31661
NN-1		Charcoal Poly	31872
OO-1		Deep Burgundy Poly	50567**
PP-1		Vermillion	71006**
TT-1		Fawn	21760
UU-1		Cocoa Poly	50543
WW-1		Satin	8218

* Dodge Only
** Dart Only

COLOR COMBINATIONS

CC-1, DD-1, etc. - Single tone

CD-2 - Standard two-tone roof only

CD-4 - With roof cantilever

CD-5 - Sweep cantilever

1961 DODGE

AA-1	D,DA	Bamboo	81234
BB-1	D,DA,LA	Midnight	9000
CC-1	D,DA,LA	Glacier Blue	12216
DD-1	D,DA,LA	Marlin Blue Poly	12272
FF-1	D,DA,LA	Spring Green	42733
GG-1	D,DA,LA	Frosted Mint Poly	42732
HH-1	D,DA	Cactus Poly	42540
JJ-1	D,DA	Turquoise	12273
KK-1	D,DA	Nassau Green Poly	12127
LL-1	D,DA,LA	Silver Gray Poly	31746
PP-1	D,DA,LA	Vermilion	71203
SS-1	D,DA	Rose Mist Poly	71147
UU-1	D,DA	Aztec Gold Poly	21949
WW-1	D,DA,LA	Snow	8218
YY-1	D,DA	Buckskin	21905
ZZ-1	D,DA	Roman Bronze Poly	21927

D-Dodge, DA-Dart, LA-Lancer

COLOR COMBINATIONS

STYLE NO. 1 - Single tone.

STYLE NO. 4 - Saddle two-tone on Lancer two door hard +op only. 1st letter indicates saddle color, 2nd letter basic body color.

STYLE NO. 2 - Dodge and Dart except two door hard top,roof only.1st letter indicates roof color, 2nd letter basic body color.

STYLE NO. 5 - Dodge and Dart two door hard top only, roof sweep. 1st letter indicates roof sweep color, 2nd letter basic body color.

1962 DODGE

AA-1	D,DA	Flax	81285
BB-1	D,DA,L,P	Onyx	9000
CC-1	D,DA,L	Powder Blue	12403
DD-1	D,DA,L,P	Medium Blue Poly	12416
EE-1	D,DA	Cobalt Blue Poly	12415
FF-1	D,DA,L	Light Green	42840
GG-1	D,DA,L,P	Glade Green Poly	42854
HH-1	D,DA	Metallic Emerald Poly	42885
MM-1	D,DA,L	Pearl Gray	32127
PP-1	D,DA,L,P	Vermilion	71203
RR-1	D,DA	Dusty Rose Poly	21966
SS-1	D,DA	Deep Cordovan Poly	22025
TT-1	D,DA,L	Buff	32074
UU-1	P	Shell Beige	22131
WW-1	D,DA,L	Polar	8293
YY-1	D,DA,L,P	Nutmeg Brown Poly	22023

D-Dodge, DA-Dart, L-Lancer, P-Polara
SINGLE TONE - AA-1, BB-1, etc.

TWO-TONES - BW-1 First letter, accent color, roof, saddle. Second letter, body color.

1963 DODGE

AA-1	D,DA,P	Turquoise Poly	12663
BB-1	D,DA,CU,P	Onyx	9000
CC-1	D,DA,CU,P	Light Blue	12469
DD-1	D,DA,P,CU	Medium Blue Poly	12625
EE-1	D,DA,P,CU	Dark Blue Poly	12658
GG-1	CU	Light Green Poly	43131
HH-1	CU	Forest Green Poly	43134
JJ-1	D,DA,P	Aqua	43001
KK-1	CU	Slate Turquoise Poly	12724
LL-1	CU	Dark Turquoise Poly	12723
MM-1	D,DA,CU,P	Ivory	32202
NN-1	D,DA,P,CU	Steel Gray Poly	32263
PP-1	D,DA,CU,P	Vermilion	71203
WW-1	D,DA,CU,P	Polar	8293
XX-1	D,DA,CU,P	Beige	22234
YY-1	D,DA,P,CU	Sandalwood Poly	22218
ZZ-1	CU	Cordovan Poly	22285

D-Dodge, DA-Dart, CU-Custom 880, P-Polara
NOTE: Silver 31746 is used on Dodge and Polara stone shields.
SINGLE TONE - AA-1, BB-1, etc.

TWO-TONES - BW-2 First letter, accent color, roof, saddle. Second letter, body color.

1964 DODGE

BB-1	D,DA,880	Black	9000
CC-1	D,DA,880	Light Blue	12655
DD-1	D,DA	Medium Blue Poly	12656
	880		12763
EE-1	D,DA	Dark Blue Poly	12657
	880		12764
JJ-1	D,DA	Aqua	12736
KK-1	D,DA	Medium Turquoise	12647
	880	Light Turquoise	12648
		Poly	
LL-1	D,DA	Dark Green Poly	12708
	880		12765
MM-1	D	Gray Poly	32263
	880		32305
PP-1	D,DA	Red	71203
SS-1	D,DA,880	Ivory	81403
TT-1	880	Dark Red Poly	71393
	D,DA	Signet Royal	71355
		Red Poly	

WW-1	D,DA,880	White	8358
XX-1	D,DA,880	Beige	22293
YY-1	D,DA	Tan Poly	22296
	880		22317
ZZ-1	D,DA	Anniversary Gold	81404
	Poly		
	880	81416	

D-Dodge, DA-Dart, 880-880
SINGLE TONE - AA-1, BB-1, etc.

TWO-TONES - BW-2 First letter, accent or roof color. Second letter, basic body color.

1965 DODGE

AA-1	Gold Poly	22461
BB-1	Black	9000,9300
CC-1	Light Blue	12894
DD-1	Medium Blue Poly	12763
EE-1	Dark Blue Poly	12896
FF-1	Pale Blue Poly	12895
GG-1	Dark Green Poly	43149
JJ-1	Light Turquoise	12901
KK-1	Medium Turquoise Poly	12897
LL-1	Dark Turquoise Poly	12765
NN-1	Pale Silver Poly	32398
PP-1	Bright Red	71433
RR-1	Beige	22441
SS-1	Ivory	81413
TT-1	Ruby Red Poly	71476
VV-1	Cordovan Poly	50673
WW-1	White	8362
XX-1	Light Tan	22440
YY-1	Medium Tan Poly	22643
ZZ-1	Pale Turquoise Poly	12898
22-1	Medium Green Poly	43287
33-1	Pink Gold Poly	22444
77-1	Pale Gold Poly	22442
88-1	Yellow	81515

TWO-TONES-First digit or letter, roof color. Second digit or letter, body color.

1966 DODGE

AA-1	Silver Poly	32398
BB-1	Black	9000,9300
CC-1	Light Blue	13037
DD-1	Medium Blue Poly	13043
EE-1	Dark Blue Poly	13040
FF-1	Light Green Poly	43414
GG-1	Dark Green Poly	43149
KK-1	Pale Medium Turquoise	12898
	Poly	

LL-1	Dark Turquoise Poly	12765
MM-1	Turbine Bronze Poly	60492
PP-1	Bright Red	71483
QQ-1	Red Poly	71476
RR-1	Yellow	81515
SS-1	Cream	81501
WW-1	White	8362
XX-1	Beige	22541
YY-1	Bronze Poly	22538
ZZ-1	Gold Poly	22511
44-1	Sandstone Poly	22443
66-1	Mauve Poly	50702
77-1	Maroon Poly	50699

1967 DODGE

AA-1	Silver Poly	32398
BB-1	Black	9000,9300
CC-1	Medium Blue Poly	13159
DD-1	Light Blue Poly	13043
EE-1	Dark Blue Poly	13040
FF-1	Light Green Poly	43547
GG-1	Dark Green Poly	43540
HH-1	Dark Copper Poly	22659
JJ-1	Chestnut Poly	22704
KK-1	Medium Turquoise Poly	13195
LL-1	Dark Turquoise Poly	13214
MM-1	Turbine Bronze Poly	60492
PP-1	Bright Red	71483
QQ-1	Dark Red Poly	71552
RR-1	Yellow	81515
SS-1	Cream	81501
TT-1	Medium Copper Poly	22706
WW-1	White	8362
XX-1	Light Tan	22701
YY-1	Medium Tan Poly	22700
ZZ-1	Gold Poly	22715
66-1	Mauve Poly	50731
88-1	Bright Blue Poly	13336

SINGLE TONE - AA-1, BB-1, 88-1, etc.

TWO-TONE - BW-2, W8-2, etc. The first digit or letter, roof color. The second digit or letter, body color.

1968 DODGE

AA-1	Silver Poly	8588
BB-1	Black	9000,9300
CC-1	Medium Blue Poly	13355
DD-1	Pale Blue Poly	13360
EE-1	Dark Blue Poly	13372
FF-1	Light Green Poly	43646
GG-1	Racing Green Poly	43649

HH-1	Light Gold	81575
JJ-1	Medium Gold Poly	22807
KK-1	Light Turquoise Poly	13195
LL-1	Med. Dark Turquoise Poly	13371
MM-1	Bronze Poly	60492
PP-1	Red	71483
QQ-1	Bright Blue Poly	13354
RR-1	Burgundy Poly	50749
SS-1	Yellow	81574
TT-1	Medium Green Poly	43647
UU-1	Light Blue Poly	13445
WW-1	White	8653
XX-1	Beige	22441
YY-1	Medium Tan Poly	22855
33-1	Charger Red	71582
66-1	Dark Green Poly	43786

1969 DODGE

A-4	Silver Poly	2016
B-3	Light Blue Poly	2018
B-5	Bright Blue Poly	2019
B-7	Medium Blue Poly	2020
B-9	Dark Blue Poly	2021
E-7	Cordovan Poly	2022
F-3	Light Green Poly	2023
F-5	Medium Green Poly	2024
F-6	Bright Green Poly	2103
F-8,66-1	Dark Green Poly	43786
L-1	Beige	22542
Q-4	Light Turquoise Poly	2028
Q-5	Bright Turquoise Poly	13534
R-4,33-1	Charger Red	71582
R-6	Red	2029
T-3	Light Bronze Poly	2030
T-5	Copper Poly	2031
T-7	Dark Brown Poly	2032
V-2	Hemi Orange	2186
W-1	White	2033
Y-2	Yellow	81574
Y-3	Cream	81575
Y-4	Gold Poly	2034
999	Orange	60436
999	Rallye Green	44032
999	Bahama Yellow	81570
X-9	Black	9300

1970 DODGE

A-4	Silver Poly	2016
A-9	Dark Gray Poly	2017
B-3	Light Blue Poly	2018
B-5	Bright Blue Poly	2019

B-7	Dark Blue Poly	2020
C-7	Plum Crazy	2210
E-5	Bright Red	2136
F-4	Light Green Poly	2133
F-6	Bright Green Poly	2103
F-8	Dark Green Poly	43786
X-9	Black	9300
J-5	Sublime	2128
J-6	Green Go	2259
K-2	Go Mango	2201
K-5	Dark Burnt Orange Poly	2135
L-1	Beige	22542
M-3	Panther Pink	2260
Q-3	Light Turquoise Poly	2127
R-6	Red	2029
R-8	Burgundy Poly	50749
T-3	Tan Poly	2131
T-6	Dark Tan Poly	2129
V-2	Hemi Orange	2186
W-1	White	2033
Y-1	Banana	2211
Y-3	Cream	81575
Y-4	Light Gold Poly	2117
Y-6	Gold Poly	2102

1971 DODGE

A-4	Light Gunmetal Poly	2314
A-8	Gunmetal Gray Poly	2315
B-2	Light Blue Poly	2304
B-5	Brite Blue Poly	2306
B-7	Dark Blue Poly	2302
C-7	Plum Crazy	2210
C-8	Indigo Poly	2305
E-5	Bright Red	2136
E-7	Burgundy Poly	2321
F-3	Medium Green Poly	2316
F-7	Dark Green Poly	2317
J-3	Willow Green	2383
J-4	Lime Green Poly	2319
J-6	Green Go	2259
K-6	Dark Bronze Poly	2312
L-5	Butterscotch	2325
Q-5	Turquoise Poly	2301
T-2	Tan Poly	2313
T-8	Dark Tan Poly	2309
V-2	Hemi-Orange	2186
W-1	White	2033
W-3	Brite White	2300
X-9	Black	9300
Y-1	Top-Banana Yellow	2211
Y-3	Citron Yella	2320

Y-4	Light Gold	2310
Y-7	Heritage Gold Poly	2384
Y-8	Gold Poly	2307
Y-9	Dark Gold Poly	2311

1972 DODGE

A-4	Light Gunmetal Poly	2314
A-5	Silver Frost Poly	2513
A-9	Charcoal Poly	2017
B-1	Powder Blue	2424
B-3	Blue Streak	2423
B-5	Brite Blue Poly	2306
B-7	Midnight Blue Poly	2302
B-9	Regal Blue Poly	2508
E-5	Red	2136
F-1	Mist Green	2515
F-3	Fiesta Green Poly	2316
F-5	Bright Yellow Green Poly	2461
F-7	Sherwood Green Poly	2317
F-8	Forest Green Poly	2514
JY-9	Tahitian Gold Poly	2510
L-4	Summer Sand	2427
Q-5	Turquoise Poly	2301
T-6	Doeskin Poly	2426
T-8	Dark Tan Poly	2425
V-2	Hemi-Orange	2186
W-1	White	2033
W-3	Brite White	2300
X-9	Black	9300
Y-1	Top-Banana Yellow	2211
Y-3	Citron Yella	2320
Y-4	Light Gold	2310
Y-7	Heritage Gold Poly	2384
Y-8	Gold Poly	2307
Y-9	Dark Gold Poly	2311

1973 DODGE

A-5	Dark Silver Poly	2513
B-1	Light Blue	2424
B-3	Blue Streak	2423
B-5	Bright Blue Poly	2306
B-9	Dark Blue Poly	2508
E-5	Bright Red	2136
F-1	Pale Green	2515
F-3	Light Green Poly	2316
*F-5	Bright Yellow Green Poly	2461
F-8	Dark Green Poly	2514
K-6	Bronze Poly	2312
L-4	Parchment	2427
L-6	Aztec Gold Poly	2591
Q-5	Turquoise Poly	2301

T-6	Medium Tan Poly	2426
T-8	Dark Tan Poly	2425
W-1	Eggshell White	2033
X-9	Black	9000,9300
Y-1	Top Banana	2211
Y-2	Yellow	81574
Y-3	Light Gold	2517
Y-6	Gold Poly	2509
JY-9	Dark Gold Poly	2510

1974 DODGE

A-5	Dark Silver Poly	2513
B-1	Powder Blue	2626
B-5	Lucerne Blue Poly	2627
B-8	Starlight Blue Poly	2628
E-5	Bright Red	2136
E-7	Burnished Red Poly	2321
G-2	Frosty Green Poly	2629
G-8	Deep Sherwood Poly	2631
J-6	Avocado Gold Poly	2632
L-4	Parchment	2427
L-6	Aztec Gold Poly	2591
L-8	Dark Moonstone Poly	2633
T-5	Sienna Poly	2634
T-9	Dark Chestnut Poly	2590
W-1	Eggshell White	2033
X-9	Black	9300
Y-2	Yellow	81574
Y-4	Golden Fawn	2635
Y-5	Yellow Blaze	2636
Y-6	Gold Poly	2509
Y-9	Dark Gold Poly	2510

TWO-TONES: First two digits are accent or roof color. Second two digits are basic body color.

1975 DODGE

A-2	Silver Cloud Poly	2734
A-5	Silver Frost Poly	2513
B-1	Powder Blue	2626
B-2	Astral Blue Poly	2735
B-5	Lucerne Blue Poly	2627
B-8	Starlight Blue Poly	2628
E-5	Bright Red	2136
E-9	Vintage Red Poly	2736
G-2	Frosty Green Poly	2629
G-8	Deep Sherwood Poly	2631
J-2	Platinum Poly	2730
J-6	Avocado Gold Poly	2632
K-3	Bittersweet Poly	2740
L-4	Parchment	2427

L-5	Moondust Poly	2737
L-6	Aztec Gold Poly	2591
R-6	Claret Red	2854
T-4	Cinnamon Poly	2741
T-5	Sienna Poly	2634
T-9	Dark Chestnut Poly	2590
W-1	Eggshell White	2033
X-9	Black	9300
Y-4	Golden Fawn	2635
Y-5	Yellow Blaze	2636
Y-6	Inca Gold Poly	2738
Y-9	Spanish Gold Poly	2739

TWO-TONES: First two digits are accent or roof color. Second two digits are basic body color.

1976 DODGE

A-2	Silver Cloud Poly	2734
A-5	Silver Frost Poly	2513
B-1	Powder Blue	2626
B-2	Astral Blue Poly	2735
B-4	Big Sky Blue	2850
B-5	Jamaican Blue Poly	2851
B-8	Starlight Blue Poly	2628
E-5	Bright Red	2136
E-8	Vintage Red Sunfire Poly	2849
E-9	Vintage Red Poly	2736
F-2	Jade Green Poly	2852
G-8	Deep Sherwood Poly	2631
G-9	Deep Sherwood Sunfire Poly	2848
J-2	Platinum Poly	2730
J-5	Tropic Green Poly	2853
K-3	Bittersweet Poly	2740
L-4	Parchment	2427
L-5	Moondust Poly	2737
R-6	Claret Red	2854
T-4	Cinnamon Poly	2741
T-9	Dark Chestnut Poly	2590
U-2	Saddle Tan	2855
U-3	Carmel Tan Poly	2856
U-6	Little Chestnut Poly	2857
V-1	Spitfire Orange	2858
W-1	Eggshell White	2033
X-9	Black	9300,9000
Y-3	Harvest Gold	2859
Y-4	Golden Fawn	2635
Y-5	Yellow Blaze	2636
Y-6	Inca Gold Poly	2738
Y-7	Taxi Yellow	81746
Y-9	Spanish Gold Poly	2739

TWO-TONES: First two digits are accent or roof color. Second two digits are basic body color.

1977 DODGE

A-2	Silver Cloud Poly	2734
A-5	Silver Frost Poly (Two-Tone)	2513
B-2	Wedgewood Blue	2934
B-3	Cadet Blue Poly	2935
B-5	French Racing Blue	2936
B-6	Regatta Blue Poly	2937
B-9	Starlight Blue "Sunfire" Poly	2938
E-5	Bright Red	2136
E-8	Vin. Red "Sunfire" Poly	2849
F-2	Jade Green Poly	2852
F-7	Forest Green "Sunfire" Poly	2939
K-6	Burn. Copper Poly	2940
L-3	Mojave Beige	2941
L-5	Moondust Poly	2737
R-6	Claret Red	2954
R-8	Russet "Sunfire" Poly	2942
T-2	Light Mocha Tan	2943
T-7	Coffee "Sunfire" Poly	2944
U-3	Carmel Tan Poly	2856
U-6	Light Chestnut Poly	2857
V-1	Spitfire Orange	2858
W-1	Eggshell White	2033
X-8	Black "Sunfire" Poly	2945
Y-1	Jasmine Yellow	2946
Y-3	Har. Gold (Two-Tone)	2859
Y-4	Golden Fawn	2635
Y-5	Yellow Blaze	2636
Y-6	Inca Gold Poly	2738
*Y-7	Taxi Yellow	81746
Y-9	Spanish Gold Poly	2739

TWO-TONES: First two digits are accent or roof color. Second two digits are basic body color.

1977 DODGE DIPLOMAT EXTERIOR COLORS

RA1	Dove Gray	3015
RA2	Pewter Grey Poly	3016
RA9	Charcoal Grey "Sunfire" Poly	3017
RF3	Mint Green Poly	3019
RF9	Augusta Green "Sunfire" Poly	3018
RR7	Tapestry Red "Sunfire" Poly	3020
RT9	Sable "Sunfire" Poly	3014
RY3	Classic Cream	3021

1978 DODGE

DX9	Low Luster Black	9440
	Accent Color on Omni Only	
EW1	Eggshell White	2033
EY7	Taxi Yellow	81746
JA5	Silver Frost (Two-Tone)	2513
KY4	Golden Fawn	2635
KY5	Yellow Blaze	2636
LY9	Spanish Gold Poly	2739
MU9	Caramel Tan Poly	2856
MV1	Spitfire Orange	2858
PB2	Wedgewood Blue	2934
PB3	Cadet Blue Poly	2935
PB6	Regatta Blue Poly	2937
PB9	Starlight Blue "Sunfire" Poly	2938
PT2	Light Mocha Tan	2943
PY1	Jasmine Yellow	2946
RA1	Dove Grey	3015
RA2	Pewter Grey Poly	3016
RA9	Charcoal Grey "Sunfire" Poly	3017
RF3	Mint Green Poly	3019
RF9	Aug. Green "Sunfire" Poly	3018
RJ3	Citron Poly	3066
RK2	Sunrise Orange	3067
RR4	Brite Canyon Red	3068
RR7	Tap. Red "Sunfire" Poly	3020
RR9	Crim. Red "Sunfire" Poly (Two-Tone)	3064
RT9	Sable "Sunfire" Poly	3014
RY3	Classic Cream	3021
TX9	Black	9000,9300

Chapter 7   EDSEL 1958-1960

PAINT CODE	COLOR	PPG CODE
1958 EDSEL		
A01	Black	9000
B50	Silver Gray Poly	31589
C40	Ember Red	70794
D95	Turquoise	41991
E	Snow White(Ranger Pacer)	8103
E	Frost White(Corsair Citation)	8150
F25	Powder Blue	11687
G26	Horizon Blue	11703
H27	Royal Blue Poly	11693
J10	Ice Green	42165
K11	Spring Green	42164
L12	Spruce Green Poly	42168
M60	Charcoal Brown Poly	21469
N	Driftwood	31346
Q70	Jonquil Yellow	80840
R	Sunset Coral	70793
T85	Chalk Pink	70632
U80	Copper Poly	21473
X90	Durez Gold Poly	21463
1959 EDSEL		
A	Jet Black	9000
B	Moonrise Gray	31316
C	Durex Gold Poly	21705
D	Redwood Poly	21738
E	Snow White	8103
F	President Red	70972
G	Talisman Red	70850
H	Desert Tan	21690
J	Velvet Maroon	70934
K	Platinum Gray Poly	31813
L	Star Blue Poly	12032
M	Jet Stream Blue	11467
N	Light Aqua	11978
P	Blue Aqua	1977
Q	Petal Yellow	80840
R	Mist Green	42433
S	Jadeglint Green Poly	42510
1960 EDSEL		
A	Black Velvet	9000
C	Turquoise	42434
E	Cadet Blue Poly	12236
F	Hawaiian Blue	12147
H	Alaskan Gold Poly	21857
J	Regal Red	71054
K	Turquoise Poly	12238
M	Polar White	8238
N	Sahara Beige	21828
Q	Lilac Poly	50570
R	Buttercup Yellow	81052
T	Sherwood Green Poly	42344
U	Bronze Rose Poly	21849
W	Seafoam Green	42634
Z	Cloud Silver Poly	31991

COLOR COLOR	PPG		
CODE	**CODE**		
1941 FORD		Blue Gray Poly	30409
Black	9000	Tucson Tan	20384
Cotswold Gray Poly	30059	Maize Yellow	80274
Lochaven Green(Original)	40054	Pheasant Red	50102
Lochaven Green(Weathered)	40121	Midland Maroon Poly	50121
Cayuga Blue	10059	Shoal Green Gray Poly	30499
Mayfair Maroon	50012	Strato Blue	10403
Palisade Gray	30034	Black	9000
Capri Blue Poly	10060		
Florentine Blue	10029	**1949 FORD**	
Seminole Brown	20049	Black	9000
Sheffield Gray Poly	30075	Colony Blue	10428
Conestoga Tan Poly	20050	Gunmetal Gray Poly	30481
		Seamist Green	40575
1942 FORD		Midland Maroon Poly	50121
Florentine Blue	10029	Miami Cream	80356
New Castle Gray	30037	Bayview Blue Polv	10461
Niles Blue Green	40026	Birch Gray	30153
Fathom Blue	10096	Meadow Green	40481
Moselle Maroon	50013	Fez Red	50104
Village Green	40101		
Phoebe Gray Poly	30038	**1950 FORD**	
Black	9000	Cambridge Maroon Poly	50165
		Osage Green Poly	40859
1946 FORD		Dover Gray	30685
Greenfield Green	40188	Sportsman Green	40858
Light Moonstone Gray	30191	Silvertone Gray	30777
Navy Blue No. 1	10177	Palisade Green	40809
Navy Blue No. 2	10263	Sunland Beige	20489
Botsford Blue Gray	10176	Hawthorne Green Poly	40810
Modern Blue	10274	Matador Red Poly	50177
Dynamic Maroon	50041	Sheridan Blue	10428
Greenfield Green	40185	Bimini Blue Poly	10461
Dark Slate Gray Poly	30190	Black	9000
Silver Sand Poly	20143	Coronation Red Poly	70351
Willow Green	40184	Hawaiian Bronze Poly	20704
Black	9000	Wagon Tan	20453
		Casino Cream	80492
1947-48 FORD			
Rotunda Gray	30414	**1951 FORD**	
Barcelona Blue	10361	Culver Blue Poly	10756
Monsoon Maroon	50035	Coral Flame	70358
Parrot Green Poly	40455	Alpine Blue	10757
Taffy Tan	20375	Mexacali Maroon Poly	50214
Glade Green	40457	Sea Island Green Gray	30851
Feather Gray	30046	Sherman BLue	10428
		Hawthorne Green Poly	40810
		Silvertone Gray	30777

Sportsman Green	40858
Black	9000
Greenbrier Green	41046
Hawaiian Bronze Poly	20704
Dark Brown	20666
Sandpiper Beige	20789
Carnival Red Poly	50245

TWO-TONES:

U 80585	U 9000	U 20789
L 9000	L 30851	L 20704
U 41046	U 20704	U 9000
L 9000	L 20666	L 40858
U 30777	U 30851	
L 10757	L 41046	

1952 FORD

Black	9000
Coral Flame Red	70358/ 70393
Alpine Blue	10757
Sheridan Blue	10428
Woodsmoke Gray	30175
Shannon Green Poly	40360
Carnival Red Poly	50245
Sandpiper Beige	20789
Meadowbrook Green	40461
Sungate Ivory	80090
Glenmist Green	40438
Hawaiian Bronze	20296

TWO-TONES:

U 20789	U 80090	U 20789
L 50245	L 10757	L 20296
U 40438	U 20296	U 10428
L 40360	L 80090	L 10757
U 80090	U 80090	
L 40461	L 9000	

1953 FORD

Black	9000
Sheridan Blue	10428
Woodsmoke Gray	30175
Carnival Red Poly	50245
Sandpiper Beige	20789
Sungate Ivory	80090
Glacier Blue	10911
Fern Mist Green	41267
Polynesian Bronze Poly	20893
Timberline Green Poly	41272
Flamingo Red	70415

Seafoam Green	41269
Coral Flame	70393
Cascade Green	41350

TWO-TONES:

U 80090	U 41272	U 10428
L 41267	L 41269	L 80090
U 80090	U 10428	U 20789
L 70415	L 10911	L 20893
U 80090	U 80090	U 20893
L 50245	L 9000	L 20789
U 50245	U 41267	U 80090
L 80090	L 80090	L 10428
U 20789	U 80090	
L 50245	L 10911	

1954 FORD

A	Black	9000
B	Sheridan Blue	10428
C	Cadet Blue Poly	11063
E	Dovetone Gray	31103
F	Highland Green Poly	41352
G	Killarney Green Poly	41465
H	Sea Haze Green	41463
J	Lancer Maroon Poly	50371
L	Sandalwood Tan	20979
M	Sandstone White	20981
N	Torch Red	70393
O	Glacier Blue	10911
R	Cameo Coral	70455
T	Skyhaze Green (Blue)	11115
V	Goldenrod	80653
Z	Sierra Brown Poly	21105

TWO-TONES:

EC	U 11063	BD	U 10911	RM	U 20981
	L 31103		L 10428		L 70455
GH	U 41463	MN	U 70393	HF	U 41352
	L 41465		L 20981		L 41463
LJ	U 50371	TM	U 20981	JL	U 20979
	L 20979		L 11115		L 50371
MC	U 11063	GM	U 20981	NA	U 9000
	L 20981		L 41465		L 70455
CM	U 20981	AM	U 20981	VM	U 20981
	L 11063	L	9000		L 80653
MG	U 41465	MR	U 70455	ZL	U 20979
	L 20981		L 20981		L 21105

1955 FORD

A	Raven Black	9000

B	Banner Blue Poly	11232
C	Aquatone Blue	11114
D	Waterfall Blue	11250
E	Snowshoe White	8097
F	Pinetree Green Poly	41635
G	Sea Sprite Green	41669
H	Neptune Green(Blue)	11249
K	Buckskin Brown	21131
M	Regency Purple Poly	50390
R	Torch Red	70540
T	Thunderbird Blue	11115
V	Goldenrod Yellow	80653
W	Tropical Rose	50397
X	Regatta Blue	11383
Y	Mountain Green	41804
Z	Coral Mist	70608

TWO-TONES:

KE	Upper	8097	CE	Upper		8097
	Lower	21131		Lower		11114
RE	Upper	8097	GE	Upper		8097
	Lower	70540		Lower		41669
GE	Upper	8097	HF	Upper		41635
	Lower	11249		Lower		11249
HF	Upper	41635	RE	Upper		8097
	Lower	11249		Lower		70540
CE	Upper	8097	KE	Upper		8097
	Lower	11114		Lower		21131
DB	Upper	11232				
	Lower	11250				

STATION WAGON & VICTORIA:

VE	Upper	8097
	Lower	80653

VICTORIA:

VA	Upper	9000	AE	Upper	8097
	Lower	80653		Lower	9000
ME	Upper	8097	EC	Upper	11114
	Lower	50390		Lower	8097
WE	Upper	8087			
	Lower	50307			

"UNIQUE" TWO-TONES:

ER	8097	EK	8097	WE	50397		
	70540		21131		8097		
EG	8097	EA	8097	AE	9000		
	41669		9000		8097		
EC	8097	CE	11114	CE	11114		
	11114		8097		8097		
AV	9000	EV	8097	VE	80653		

	80653		80653			8097	
EM	8097	RE	70540	CD		11114	
	50390		8097			11250	
EW	8097	GE	41669	DB	1	1250	
	50397		8097			11232	
FH	41635	ME	50390	HF		11249	
	11249		8097			41635	

SPRING TWO-TONES:

Upper	8097		Upper	8097
Lower	70608		Lower	11383
Upper	70608		Upper	11383
Lower	8087		Lower	8097
Upper	8097		Upper	11383
Lower	41804		Lower	11250
Upper	41804			
Lower	8097			

1956 FORD

A	Raven Black	9000
B	Nocturne Blue Poly	11423
C	Bermuda Blue	11414
D	Diamond Blue	11403
E	Colonial White	8103
F	Pine Ridge Green Poly	41837
G	Meadowmist Green	41839
H	Platinum Gray	31316
J	Buckskin Tan	21215
K	Fiesta Red	70618
L	Peacock Blue	11415
M	Goldenglow Yellow	80788
N	Mandarin Orange	60211
V	Berkshire Green	41979
W	Springmist Green	41980
Y	Sunset Coral	70706

1956 THUNDERBIRD

A	Raven Black	9000
E	Colonial Whlte	8103
J	Buckskin Tan	21215
K	Fiesta Red	70618
L	Peacock Blue	11415
T	Gray Poly	31386
Z	Sage Green	41972
	Navajo Gray	31443

TWO-TONES, CONVENTIONAL:

AE	U	8103	LE	U	8103	HE	U		8103
	L	9000		L	11415		L		31316
KE	U	8103	CE	U	8103	AK	U		70618
	L	70618		L	11414		L		9000

FC	U	41839	DE	U	8103	ME	U	8103
	L	41837		L	11403		L	80788
GE	U	8103	GF	U	41837	JE	U	8103
	L	41839		L	41839		L	21215

U-Upper Body, L-Lower Body
BCMR - Below Crash Mldg. & Roof
CMB - Crash Mldg. to Belt
BBMR - Below Belt Mldg. & Roof
BMDM - Belt Mldg. to Drip Mldg.
BCMR - Below Crash Mldg. & Roof
CMDM - Crash Mldg. to Drip Mldg.

STYLE-TONES:

AE	BCMR	9000	NE	BCMR	60211
	CMB	8103		CMB	8103
HE	BCMR	31316	JE	BCMR	21215
	CMB	8103		CMB	8103
ME	BCMR	80788	DC	BCMR	11403
	CMB	8103		CMB	11414
AK	BCMR	9000	ED	BCMR	8103
	CMB	70618		CMB	11403
KE	BCMR	70618	GF	BCMR	41839
	CMB	8103		CMB	41837
FG	BCMR	41837	KA	BCMR	70618
	CMB	41839		CMB	9000
GE	BCMR	41839	EK	BCMR	8103
	CMB	8103		CMB	70618
LE	BCMR	11415	MA	BCMR	80788
	CMB	8103		CMB	9000
CD	BCMR	11414	EA	BCMR	8103
	CMB	11403		CMB	9000
CE	BCMR	11414	EN	BCMR	8103
	CMB	8103		CMB	60211
DE	BCMR	11403			
	CMB	8103			

CONVENTIONAL TWO-TONES, STATION-WAGON:

CE	BBMR	11414	AE	BBMR	9000
	BMDM	8103		BMDM	8103
GE	BBMR	41839	CD	BBMR	11414
	BMDM	8103		Parklane BMDM	11403
KE	BBMR	70618		only	
	BMDM	8103	FG	BBMR	41837
JE	BBMR	21215		Parklane BMDM	41839
	BMDM	8103		only	
ME	BBMR	80788	JE	BBMR	21215
	BBDM	8103		Parklane BMDM	8103
HE	BBMR	31316		only	
	BMDM	8103			

STYLE TWO-TONES, STATION WAGON:

CE	BCMR	11414	ME	BCMR	80788
	CMDM	8103		CMDM	8103
GE	BCMR	41839	AE	BCMR	9000
	CMDM	8103		CMDM	8103
HE	BCMR	31316	JE	BCMR	21215
	CMDM	8103		CMDM	8103
KE	BCMR	70618			
	CMDM	8103			

1957 FORD

A	Raven Blach	9000
C	Dresden Blue	11602
D	Silver Mocha Poly	21469
E	Colonial White	8103
F	Starmist Blue	11547
G	Cumberland Green	42027
H	Gunmetal Gray Poly	31578
J	Willow Green	42025
K	Silver Mocha Poly Dk.	21376
L	Doeskin Tan	21331
N	Gunmetal Gray Poly	31464
Q	Thunderbird Bronze Poly	50469*
T	Woodsmoke Gray	31415
V	Flame Red	70707
V	Berkshire Green	41979
W	Springmist Green	41980
X	Thunderbird Dusk Rose	50470*
Y	Sunset Coral	70706
Y	Inca Gold	80859
Z	Coral Sand	70708

*Thunderbird Only

TWO-TONES: In color combinations the first letter indicates lower color, the second letter generally upper or middle color.

1958 FORD

A	Raven Black	9000
C	Desert Beige	21484
D	Palamino Tan	21485
E	Colonial White	8103
F	Silvertone Green Poly	41293
G	Sun Gold	80948
H	Gunmetal Gray Poly	31578
J	Bali Bronze Poly	21528
L	Azure Blue	11692
M	Gulfstream Blue	42121
N	Seaspray Green	42171
R	Torch Red	70801
T	Silvertone Blue Poly	11716

1958 THUNDERBIRD

A	Black	9000
B	Winterset White	8050
I	Grenadier Red Poly	70879
K	Everglade Green Poly	42132
M	Gulfstream Blue	42121
O	Platinum Poly	31608
P	Palamino Tan	21485
V	Casino Cream	80918
W	Cameo Rose	70789
X	Cascade Green	42086
Y	Monarch Blue Poly	11661
Z	Regatta Blue	11662

1959 FORD

A	Raven Black	9000
C	Wedgewood Blue	11839
D	Indian Turquoise	42256
E	Colonial White	8103
F	Fawn Tan	21641
G	April Green	42329
H	Tahitian Bronze Poly	21712
J	Surf Blue Poly	11979
Q	Sherwood Green Poly	42461
R	Torch Red	70540
T	Geranium	60321
Y	Inca Gold	80859
Z	Gunsmoke Gray Poly	70833
1	Grenadier Red Poly	70879

TWO-TONES: On styletone color combinations the first letter indicates the color below the body side moulding and the roof, the second letter indicates the middle color. On conventional two-tone combinations the first letter indicates the lower color, the second roof color.

1959 FORD THUNDERBIRD

A-01	Raven Black	9000
C	Baltic Blue	11990
D	Indian Turquoise	42256
E-07	Colonial White	8103
F-60	Hickory Tan	21400
G-15	Glacier Green	42396
H	Tahitian Bronze Poly	21712
J-28	Steel Blue Poly	11757
9-92	Sandstone Poly	21648
L-80	Diamond Blue	11683
M-66	Doeskin Beige	21517
N	Starlet Blue Poly	11930
9-17	Sea Reef Green Poly	42399
R	Brandy Wine	70980

T-83	Flamingo	70883
U	Cordovan Poly	21716
V-75	Casino Cream	80918
W	Tamarack Green Poly	42468
Z-55	Platinum	31608

1960 FORD

A	Raven Black	9000(F,FA,TB)
B	Kingston Blue Poly	12144(TB)
C	Aquamarine	42434(F,TB)
E	Acapulco Blue Poly	12164(TB)
F	Surf Foam Blue	12147(F,FA,TB)
G	Yosemite Yellow	81174(F)
H	Beechwood Brown Poly	21857(F,TB)
J	Monte Carlo Red	71054(F,FA,TB)
K	Sultana Turquoise	12238(F,FA,TB)
M	Corinthian White	8238(F,TB)
N	Diamond Blue	11683(TB)
Q	Orchid Gray Poly	50570(F)
R	Moroccan Ivory	81052(TB)
S	Briar Cliffe Green Poly	42663(TB)
T	Meadowvale Green Poly	42344(F,FA,TB)
U	Springdale Rose Poly	21849(TB)
V	Palm Spring Rose	71055(TB)
W	Adriatic Green	42634(F,FA,TB)
X	Royal Burgundy Poly	50538(TB)
Y	Gunpowder Gray Poly	31956(TB)
Z	Platinum Poly	31991(F,FA,TB)

F-Ford, FA-Falcon, TB-Thunderbird

TWO-TONES: the first letter indicates lower color, the second, upper color.

1961 FORD

A	Raven Black	9000(F,FA,TB)
C	Aquamarine	45434(F,FA,TB)
D	Starlight Blue	12355(F,FA,TB)
E	Laurel Green Poly	42820(F,FA,TB)
F	Desert Gold	81249(F,TB)
H	Chesapeake Blue Poly	12366(F,FA,TB)
J	Monte Carlo Red	71054(F,FA,TB)
K	Algiers Bronze Poly	21981(F,FA)
N	Diamond Blue	11683(TB)
P	Nautilus Gray Poly	32081(TB)
Q	Silver Gray Poly	32089(F,FA,TB)
R	Cambridge Blue Poly	12361(F,FA,TB)
S	Mint Green	42812(F,FA,TB)
T	Honey Beiee	81224(TB)
V	Palm Springs Rose	71055
W	Garden Turquoise Poly	12359(F,FA,TB)
X	Heritage Burgundy Poly	50593(TB)

Y   Mahogany Poly   21959(TB)
Z   Fieldstone Tan Poly   21958(TB)
F-Ford, FA-Falcon, TB-Thunderbird

TWO-TONES: the first letter indicates lower color, the second, upper color.

1962 FORD

A	Raven Black	9000(F,FA,FR,TB)
B	Peacock Blue	12676(F,FA,FR)
D	Patrician Green Poly	11921(TB)
D	Ming Green Poly	12489(F,FA,FR)
E	Acapulco Blue Poly	12164(TB)
F	Skymist Blue	12147(TB)
F	Baffin Blue	12488(F,FA,FR)
G	Silver Mist Poly	12497(TB)
H	Caspian Blue Poly	12597(TB)
H	Oxford Blue Poly	12496(F,FA,FR)
I	Castilian Gold Poly	31842
I	Castilian Gold Poly	32277(F,FA,FR)
J	Rangoon Red	71243(F,FA, FR,TB)
K	Chalfonte Blue	42919(TB)
L	Sahara Rose	71239(TB)
M	Corinthian White	8238(F,FA,FR,TB)
N	Diamond Blue	11683(TB)
P	Silver Moss Poly	42929(F,FA,FR)
Q	Silver Gray Poly	32089(F,FA,FR)
R	Tucson Yellow	81324(F,FA, FR,TB)
S	Cascade Green Poly	42925(TB)
T	Sandshell Beige	22110(F,FA, FR,TB)
U	Deep Sea Blue Poly	12493(TB)
V	Chestnut Poly	60392(F,FR,TB)
X	Heritage Burgundy Poly	50593(TB)
Z	Fieldstone Tan Poly	21958(F,FA, FR,TB)

F-Ford, FA-Falcon, FR-Fairlane, TB-Thunderbird

TWO-TONES: the first letter indicates lower color, the second, upper color.

1963 FORD

A	Raven Black	9000(F,FA,FR,TB)
B	Peacock Blue	12676(F,FA,FR)
D	Patrician Green Poly	11921(TB)
D	Ming Green Poly	12489(F,FA,FR)
E	Acapulco Blue Poly	12164(TB)
E	Viking Blue Poly	12494(F,FA,FR)
G	Silver Mink Poly	12497(TB)

H	Caspian Blue Poly	12547(TB)
H	Oxford Blue Poly	12496(F,FA,FR)
I	Champagne Poly	32277(F,FA, FR,TB)
J	Rangoon Red	71243(F,FA, FR,TB)
K	Chalfonte Blue	42919(TB)
L	Sahara Rose	71239(TB)
M	Corinthian White	8238(F,FA,FR,TB)
N	Diamond Blue	11683(TB)
O	Green Mist Poly	43082(TB)
P	Silver Moss Poly	42929(F,FA,FR)
R	Tucson Yellow	81324(TB)
S	Cascade Green Poly	42925(TB)
T	Sandshell Beige	22110(F,FA, FR,TB)
U	Deep Sea Blue Poly	12493(TB)
V	Chestnut Poly	60392(F,FA,TB)
W	Rose Beige Poly	71322(F,FA, FR,TB)
X	Heritage Burgundy Poly	50593(F,FA, FR,TB)
Y	Glacier Blue	12617(F,FA,FR)
Z	Fieldstone Tan Poly	21958(TB)

F-Ford, FA-Falcon, FR-Fairlane, TB-Thunderbird

TWO-TONES: the first letter indicates lower color, the second, upper color.

1964 FORD

A	Raven Black	9000(F,FR,FA,G,M)
B	Pagoda Green	12851(F,G,M)
D	Dynasty Green Poly	12853(F,FR,FA,G,M)
F	Guardsman Blue Poly	12832(F,FR,FA,G.M)
G	Prairie Tan	22230(F,FR,FA,G)
H	Caspian Blue Poly	12752(M)
J	Rangoon Red	71243(F,FR,FA,G,M)
K	Silver Smoke Gray Poly	32377(F,FR,FA,G,M)
M	Wimbledon White	8378(F,FR,FA,G,M)
P	Prairie Bronze Poly	22438(F,FR,FA,G,M)
R	Phoenician Yellow	81444(F,FR,G,M)
S	Cascade Green Poly	42925(F,FR,FA,G,M)
T	Navajo Beige	22249(F,G)
V	Sunlight Yellow	81467(F,FR,FA,G,M)
X	Vintage Burgundy Poly	50657(F,FR,FA,G,M)
Y	Skylight Blue	12850(F,FR,FA,G,M)
Z	Chantilly Beige Poly	22393(F,FR,FA,G,M)
3	Poppy Red	60449(M)
	Pace Car White	DDL-8321(M)

F-Ford, FR-Fairlane, FA-Falcon, G-Galaxie, M-Mustang

TWO-TONES: the first letter indicates lower color, the second, upper color.

1965 FORD

A	Raven Black	9000,9300
C	Honey Gold Poly	22581
D	Dynasty Green Poly	12853
F	Arcadian Blue	12854
H	Caspian Blue Poly	12547
I	Champagne Beige Poly	22436
J	Rangoon Red	71243
K	Silver Smoke Gray Poly	32377
M	Wimbledon White	8378
O	Tropical Turquoise	12852
p	Prairie Bronze Poly	22438
R	Ivy Green Poly	43337
X	Vintage Burgundy Poly	50657
Y	Silver Blue Poly	12164
3	Poppy Red Poly	60449
5	Twilight Turquoise Poly	12893
7	Phoenician Yellow	81444
8	Springtime Yellow	81510

TWO-TONES: the first letter indicates lower color, the second, upper color.

1966 FORD

A	Raven Black	9000,9300
F	Arcadian Blue	12854
H	Sahara Beige	22528
K	Nightmist Blue Poly	13076
M	Wimbledon White	8378
P	Antique Bronze Poly	22603
R	Ivy Green Poly	43408
T	Candyapple Red	71528
U	Tahoe Turquoise Poly	12745
V	Emberglo Poly	22610
X	Vintage Burgundy Poly	50669
Y	Silver Blue Poly	13045
Z	Sauterne Gold Poly	43433
4	Silver Frost Poly	32520
5	Signal Flare Red	71529
8	Springtime Yellow	81510

TWO-TONES: the first letter indicates lower color, the second, upper color.

1967 FORD

A	Raven Black	9000,9300
B	Frost Turquoise	12876
D	Acapulco Blue Poly	13357
F	Arcadian Blue	22711
H	Diamond Green	43575
I	Lime Gold Poly	43576
K	Nightmist Blue Poly	13076
M	Wimbledon White	8378
	Wimbledon White #2	8734
N	Diamond Blue	11683
Q	Brittany Blue Poly	12843
S	Dusk Rose	50470
T	Candyapple Red	71528
V	Burnt Amber Poly	22749
W	Clearwater Aqua Poly	13073
X	Vintage Burgundy Poly	50669
Y	Dark Moss Green Poly	43567
Z	Sauterne Gold Poly	43433
4	Silver Frost Poly	32520
5	Dark Gray Poly	32600
6	Pebble Beige	22249
8	Springtime Yellow	81510
	Playboy Pink	71617
	Anniversary Gold Poly #1	22157
	Anniversary Gold Poly #2	23072
	Anniversary Gold Poly #3	23073
	Columbine Blue	11666
	Aspen Gold	81434
	Blue Bonnet	13660
	Timberline Green	42750
	Lavender	50802
	Red	71697

TWO-TONES: the first letter indicates lower color, the second, upper color.

1968 FORD

A	Raven Black	9000,9300
B	Royal Maroon	50746
D	Acapulco Blue Poly	13357
F	Gulfstream Aqua Poly	13329
1	Lime Gold Poly	43576
M	Wimbledon White	8378
M	Wimbledon White #2	8734
N	Diamond Blue	11683
O	Seafoam Green	43529
Q	Brittany Blue Poly	13619
R	Highland Green Poly	43644
T	Candyapple Red	71528
U	Tahoe Turquoise Poly	12745
W	Meadowlark Yellow	81584
X	Presidential Blue Poly	13356
Y	Sunlit Gold Poly	22833
5	Low Gloss Black	9295

	(Mustang hood & cowl)	
6	Pebble Beige	22249

TWO-TONES: the first letter indicates lower color, the second, upper color.

1969 FORD

A	Raven Black	9000,9300
B	Royal Maroon	50746
C	Black Jade Poly	2037
D	Pastel Gray	2038
E	Aztec Aqua	2039
F	Gulfstream Aqua Poly	13329
I	Lime Gold Poly	2054
K	Freudian Gilt Poly	2156
M	Wimbledon White	8378
	Wimbledon White #2	8734
P	Winter Blue Poly	2042
Q	Brittany Blue Poly	13619
S	Champagne Gold Poly	2044
T	Candy Apple Red	71528
W	Meadowlark Yellow	81584
X	Presidential Blue Poly	2056
Y	Indian Fire Poly	2046
	Original Cinnamon Poly	
2	New Lime	2047
	Thanks Vermillion	2151
3	Calypso Coral	60449
4	Silver Jade Poly	2048
5	Sage Bronze	22749
6	Acapulco Blue Poly	13357
	Hulla Blue Poly	
7	Anti-Establish Mint	2149
8	Dresden Blue	2050
9	Yellow	2052

TWO-TONES: the first letter indicates lower color, the second,upper color.

1970 FORD

A	Raven Black. History Onvx	9000.9300
B	Dark Maroon	2150
C	Dark Ivy Green Poly	2146
	Bring'em Back Olive Poly	
D	Bright Yellow,	2214
	Last Stand Custer	
F	Dk. Bright Aqua Poly	13329
	Young Turquoise Poly	
G	Medium Lime Poly	2152
J	Grabber Blue	2230
K	Bright Gold Poly	2156

	Freudian Gilt Poly	
M	Wimbledon White,	8378
	Knight White	
N	Diamond Blue, Baby	11683
	Blue Ice	
O	Original Cinnamon Poly	2046
P	Med. Ivy Green Poly	2147
	Three Putt Green Poly	
Q	Medium Blue Poly	2138
	There She Blue Poly	
S	Champagne Gold Poly	2044
	Good Clean Fawn Poly	
T	Candy Apple Red	71528
	Counter Revolutionary Red	
U	Grabber Red	2232
W	Yellow	2157
X	Dark Blue	2139
Z	Grabber Green	2231
1	Calypso Coral	60449
2	New Lime	2047
5	Ginger Poly, Medium	2160
	Brown Poly	
6	Acapulco Blue Poly	13357
	Hulla Blue Poly	
7	Anti Establish Mint Poly	2149
8	Morning Gold	2043
9	Pastel Yellow	81584

TWO-TONES: the first letter indicates lower color, the second, upper color.

1971 FORD

A	Raven Black, Model T Black	9000,9300
B	Maroon Poly	2295
C	Dark Green Poly	2291
D	Grabber Yellow,	2214
	Pinto Yellow	
F	Medium Blue Poly	2404
H	Light Green	2290
I	Grabber Lime, Pinto Lime	2294
J	Grabber Blue, Pinto Blue	2230
M	Wimbledon White	8378
N	Pastel Blue	11683
O	Light Yellow Gold	2298
P	Medium Green Poly	2289
Q	Winter Blue Poly	2372
S	Gray Gold Poly	2326
T	Candy Apple Red	71528
V	Light Pewter Poly	2287
W	Yellow	2157
X	Dark Blue Poly	13076

Y	Dark Blue Poly	2405
Z	Grabber Green Poly	2293
2	Medium Bright Yellow	2414
3	Bright Red, Pinto Red	2296
4	Medium Lime	2385
5	Medium Brown Poly	2371
6	Acapulco Blue Poly	13357
7	Maroon	50746
8	Morning Gold	2043
9	Anti-Establish Mint Poly	2149

TWO-TONES: the first letter indicates lower color, the second, upper color.

1972 FORD

1-A	Light Gray Poly	2409
1-C	Black	9000,9300
2-A	Medium Coral	60449
2-B	Bright Red	2296
2-E	Red	71528
2-J-7	Maroon	50746
3-B	Light Blue	2403
3-D,F	Medium Blue Poly	2404
3-F	Grabber Blue	2230
3-H,Y	Dark Blue Poly	2405
3-J	Bright Blue Poly	13357
3-K	Blue Glow Poly	2499
4-B	Bright Green Gold Poly	2406
4-C	Ivy Bronze Poly	2362
4-E,4	Bright Lime, Medium Lime	2385
4-F	Medium Lime Poly	2412
4-P	Medium Green Poly	2289
4-Q	Dark Green Poly	2291
4-S	Light Green	2419
5-A	Light Pewter Poly	2287
5-H	Medium Brown Poly	2371
5-H	Ginger Poly	2482
5-J	Medium Ginger Bronze Poly	2363
5-N	Medium Orange Poly	2516
6-B	Light Goldenrod	2298
6-C	Medium Yelloe Gold	2299
6-D	Yellow	2157
6-E-2	Medium Bright Yellow	2414
6-F	Bright Yellow Gold Poly	2415
6-J	Gray Gold Poly	2326
9-A	White	8378
9-C	Creamy White	2512

TWO-TONES: the first letter indicates lower color, the second, upper color.

1973 FORD

1-C	Black	9000,9300
2-B	Bright Red	2296
2-C	Red Poly	2428
2-L	Bright Red Poly	2574
3-B	Light Blue	2403
3-D	Medium Blue Poly	2404
3-K	Blue Glow Poly	2499
3-M	Silver Blue Poly	2501
3-N,P	Light Grabber Blue	2611
3-Q	Pastel Blue	2613
4-B	Bright Green Gold Poly	2406
4-C	Ivy Bronze Poly	2362
4-N	Medium Aqua	2507
4-P	Medium Green Poly	2289
4-Q	Dark Green Poly	2291
4-S	Light Green	2419
5-A	Light Pewter Poly	2287
5-H	Medium Brown Poly	2482
5-J	Ginger Glow Poly	2363
5-L	Tan	2285
5-M	Medium Copper Poly	2504
5-T	Saddle Bronze Poly	2575
5-W	Orange	2576
6-B	Light Goldenrod	2298
6-C	Medium Yellow Gold	2299
6-D	Yellow	2157
6-E	Medium Bright Yellow	2414
6-F	Gold Glow Poly	2415
6-L	Medium Gold Poly	2497
9-A	White	8378
9-C	Special White	2512

In two-tone combinations the first two digits indicate lower color; the second two digits indicate upper color.

1974 FORD

1-C	Black	9000,9300
1-G	Silver Poly	2593
2-B	Bright Red	2296
2-E	Red	71528
2-M	Dark Red	2609
3-B	Light Blue	2403
3-D	Medium Blue Poly	2404
3-E	Bright Blue Poly	2610
3-G	Dark Blue Poly	2472
3-M	Silver Blue Poly	2501
3-N	Light Grabber Blue	2611
3-Q	Pastel Blue	2613
4-A	Pastel Lime	2411
4-B	Bright Green Gold Poly	2406
4-Q	Dark Green Poly	2291

4-T	Medium Ivy Bronze Poly	2666
4-V	Dark Yellow Green Poly	2614
4-W	Medium Lime Yellow	2615
5-H	Ginger Poly	2482
5-J	Ginger Glow Poly	2363
5-M	Medium Chestnut Poly	2504
5-T	Saddle Bronze Poly	2575
5-U	Tan Poly	2618
5-W	Orange	2576
6-C	Medium Goldenrod	2299
6-F	Yellow Gold Glow Rod	2415
6-M	Medium Dark Gold Poly	2621
6-N	Maize Yellow	2622
9-A	Wimbledon White	8378
9-C	White Decor	2512
9-D	White	2684

In two-tone combinations the first two digits indicate lower color; the second two digits indicate upper color.

1975 FORD

1-C	Black	9000,9300
1-G	Silver Poly	2593
1-H	Medium Slate Blue Poly	2612
2-B	Bright Red	2296
2-E	Red	71528
2-M	Dark Red	2609
2-Q	Maroon Poly	2716
3-E	Medium Bright Blue Poly	2610
3-G	Dark Blue Poly	2472
3-M	Silver Blue Glamour Poly	2501
3-Q	Pastel Blue	2613
4-T	Green Glow Poly	2666
4-V	Dark Yellow Green Poly	2614
4-Z	Light Green Gold Poly	2720
5-J	Ginger Glamour Poly	2363
5-M	Medium Copper Poly	2504
5-Q	Dark Brown Poly	2616
5-T	Saddle Bronze Poly	2575
5-U	Tan Glamour Poly	2618
5-W	Orange	2576
5-Y	Dark Copper Poly	2620
6-D	Yellow	2157
6-E	Bright Yellow	2414
6-L	Medium Gold Poly	2497
8-C	Medium Gold Poly	2813
9-D	White	2684
47	Light Green	2726

In two-tone combinations the first two digits indicate lower color; the second two digits indicate upper color.

1976 FORD

1-C	Black	9000,9300
1-G	Silver Poly	2593
1-H	Medium Slate Blue Poly	2612
1-N	Dove Gray	2847
2-A	Coral	60449
2-B	Bright Red	2296
2-M	Dark Red	2609
2-R	Bright Red	2830
3-E	Medium Bright Blue Poly	2610
3-G	Bright Dark Blue Poly	2472
3-M	Silver Blue Glow Poly	2501
3-S	Light Blue	2834
4-T	Med. Ivy Bronze Poly	2666
4-V	Dark Yellow Green Poly	2614
5-M	Medium Chestnut Poly	2504
5-Q	Dark Brown Poly	2616
5-T	Saddle Bronze Poly	2575
5-U	Tan Glow Poly	2618
6-E	Bright Yellow	2414
6-L	Medium Gold Poly	2497
6-P	Cream	2790
6-U	Tan	2836
6-V	Med. Gold Glamour Poly	2837
6-W	Light Gold	2838
7-Q	Light Aqua Poly	2887
7-R	Chartreuse	2889
9-D	White	2684
46	Dark Jade Poly	2725
47	Light Green	2726

In two-tone combinations the first two digits indicate lower color; the second two digits indicate upper color.

1977 FORD

1-C	Black	9000,9300
1-G	Silver Poly	2593
1-N	Dove Gray	2847
1-P	Med. Gray Poly (Two-Tone)	2967
2-A	Coral	60449
2-M	Dark Red	2609
2-R	Bright Red	2830
2-U	Lipstick Red	2833
2-Y	Rose Poly	2906(T)
3-G	Brt. Drk. Blue Poly	2472
3-U	Light Blue	2907
3-V	Brt. Blue Poly	2908
4-V	Drk. Yellow Green Poly	2614
5-Q	Dark Brown Poly	2616
6-E	Bright Yellow	2414
6-P	Cream	2790
6-U	Light Tan	2836

6-V	Medium Gold Poly	2837
7-H	Bright Aqua Poly	2910
7-L	Light Jade Poly	2911
7-R	Chartreuse	2889
7-S	Dark Emerald Poly	2912
7-T	Medium Emerald Poly	2913
8-G	Vista Orange	2915
8-H	Tan	2916
8-J	Med. Tan Poly	2917
8-K	Bright Saddle Poly	2918
8-W	Chamois Poly	2923
8-Y	Champagne Poly	2924
9-D	White	2684
46	Dark Jade Poly	2725
47	Light Green	2726

In two-tone combinations the first two colors indicate lower color; the second two digits indicate upper color.

1978 FORD

1-C	Black	9000,9300
1-G	Silver Poly	2593
1-N	Dove Gray	2847
1-P	Med. Gray Poly (Two-Tone)	2967
2-B	Bright Red	2830
2-M	Dark Red	2609
2-R	Bright Red	2830
3-A	Dark Midnight Blue	3035
3-E	Dia. Blue Poly	3041
3-G	Brt. Drk. Blue Poly	2472
3-U	Light Blue	2907
3-V	Bright Blue Poly	2908
5-M	Medium Chestnut Poly	2504
5-Q	Dark Brown Poly	2616
5-Y	Ember Poly	3046
6-E	Bright Yellow	2414
6-P	Cream	2790
6-W	Gold	2838
7-H	Bright Aqua Poly	2910
7-L	Medium Jade Poly	2911
7-Q	Light Aqua Poly	2887
7-W	Medium Jade	3044
8-J	Medium Tan Poly	2917
8-N	Dark Cord. Poly	2920
8-W	Chamois Poly	2923
8-Y	Champagne Poly	2924
9-D	White	2684
9-E	White(Special)	8321
21	Red	3060
34	Medium Blue	3048
46	Dark Jade Poly	2725
62	Antique Cream	3051

81	Russet Poly	3056
83	Light Chamois	305
85	Tangerine	3058
86	Pastel Beige	2999

In two-tone combinations the first two digits indicate lower color; the second two digits indicate upper color.

1979 FORD

1C	Black	9000,9300(F,L,M)
1G	Silver Poly	2593(F,L,M)
1E	Pewter Poly	33369(1979-1/2 Mustang Pace Car)
1N	Dove Gray	2847(F,L,M)
1P	Medium Gray Poly	2967(F,M)
1S	Medium Gray Poly	2930(L,V)
1Y	Silver Poly	3036(L,V)
2B	Bright Red	2296(F)
2D	Light Amethyst Poly	3183(Mark)
2H	Medium Red Poly	3168(F,M)
2J	Maroon	3175(F,M)
2P	Bright Red	3169(F,M)
2W	Light Red Poly (Rose)	5104(Mark)
3F	Medium Light Blue	3170(F,M)
3H	Light Wedgewood Blue Poly	3033(F,M)
3J	Bright Blue	3176(F,M)
3L	Dark Blue Poly	171(F,M)
3Q	Dark Blue Poly	3182(L)
4B	Dark Beryl Poly	3177(L,V)
4C	Medium Beryl Poly	3178(L,V)
4D	Dark Pine Poly	3172(F,M)
5A	Dark Champagne Polv	3038(L,V)
5C	Light Champagne Poly	3039(L)
5M	Medium Chestnut Poly	2504(F,M)
5N	Medium Dark Orange Poly	3173(F,M)
5P	Pastel Chamois	3167(F,M)
5Q	Dark Brown Poly	2616(F,M)
5R	Dark Cordovan Poly	2950(L,V)
5W	Medium Vaquero Poly	3174(F)
6L	Gold Poly	24450(Ranchero)
6P	Cream	2790(F,L,M)
6W	Light Gold	2898(F,M)
7L	Medium Jade Poly	2911(F,M)
8J	Medium Tan Poly	2917(F,M)
8N	Dark Cordovan Poly	2920(F,M)
8W	Chamois Poly	2923(T-Bird)
9D	White	2684(All)
9E	Special White	8321(F)
23	Dark Red Poly	3047(L,V)

34	Wedgewood Blue	3048(L)
38	Diamond Blue Poly	3050(L,V)
46	Dark Jade Poly	2725(F,M)
52	Light Champagne	3179(L,V)
62	Antique Cream	3051(F,M)
64	Bright Yellow	3063(F,M)
66	Jubilee Gold Poly	3054(L)
75	Medium Pine poly	3180(F,M)
76	Light Medium Pine	3181(F,M)
83	Light Chamois	3057(F,M)
85	Tangerine	3058(F,M)
88	Light Apricot Poly	3059(L)

F-Ford, L-Lincoln, M-Mercury,
V-Versailles, Mark-Mark V

1980 FORD

1C	Black	9000,9300(F,L,M)
1G	Silver Poly	2593(F,M)
IJ	Light Pewter Poly	3276(L,Mark)
1K	Medium Pewter Poly	3277(L,Mark)
1P	Medium Gray Poly	2967(F,M)
1S	Medium Gray Poly	2830(V)
1Y	Silver Poly	3036(V)
2G	Bright Bittersweet	3278(F,M)
2H	Medium Red Poly	3168(F,M)
2K	Candy Apple Red	3279(F,M,Mark)
2L	Maroon	3280(L,Mark)
3D	Dark Blue Poly	3581(F,M)
3F	Light Medium Blue	3170(F,M)
3H	Medium Blue Poly	3033(F,M)
3J	Bright Blue	3176(F,M)
3L	Dark Blue Poly	3171(F,M)
3Q	Dark Blue Poly	3182(L,Mark,V)
4C	Medium Beryl Poly	3178(V)
4E	Pastel Pine Poly	3282(F,M)
4F	Dark Pine Poly	3283(L,Mark)
4G	Pine Opolescent	3284(L,Mark)
5E	Light Fawn	3285(L,Mark)
5K	Chamois Poly	2997(L,Mark)
5R	Dark Cordovan Poly	2950(L,Mark,V)
5T	Bright Caramel	3286(F,M)
6B	Sand Poly	3287(F,M)
6D	Pastel Sand	3185(F,M)
6M	Pastel Rattan	3288(Mark)
6N	Bright Yellow	3289(F,M)
6Q	Medium Rattan	82434 (Mark(2-Tone)
7M	Dark Pine Poly	3290(F,M)
8A	Dark Chamois Poly	3291(F,M)
8B	Dark Chamois Poly	3292(L,Mark)
8D	Medium Bittersweet Poly	3293(F,M)

8E	Bittersweet Poly	3294(L,Mark,V)
8N	Dark Cordovan Poly	2920(F,M)
8W	Chamois Poly	3300(F,M)
9D	White	2684(All)
12	Light Gray	3295(F,M,V)
14	Anniversary Silver	33388(T-Bird Only)
15	Silver Poly	33402(L,Mark)
23	Dark Red Poly	3047(L,Mark,V)
27	Bright Red	3296(F,M)
38	Diamond Blue Poly	3050(L,Mark,V)
54	Dark Champagne Poly	3297(L,Mark)
56	Medium Fawn Poly	3298(L,Mark,V)
57	Light Fawn Poly	3299(L,Mark,V)

F-Ford, L-Lincoln, M-Mercury,
V-Versailles. Mark-Mark V

1981 FORD

1C	Black	9000,9300(All)
1G	Silver Poly	2593(F,M)
IJ	Light Pewter Poly	3276(L,Mark)
1K	Medium Pewter Poly	3277(L,Mark)
1P	Medium Gray Poly	2967(F,M)
1T	Light Pewter Poly	3376(F,M)
2G	Bright Bittersweet	3278(F,M)
2H	Medium Red Poly	3168(F,M)
2K	Candy Apple Red	3279(F,M,)
2Q	Maroon Poly	2716(M)
3D	Dark Blue Poly	3581(F,M)
3F	Light Medium Blue	3170(F,M)
3H	Medium Blue Poly	3033(F,M)
3L	Dark Blue Poly	3171(F,M)
3M	Bright Blue Poly	3377(F,M)
3Q	Dark Blue Poly	3182(L,Mark)
3Y	Light Blue Poly	3389(L,Mark)
4E	Pastel Pine Poly	3282(F,M)
4F	Dark Pine Poly	3283(L,Mark)
4G	Pine Opolescent	3284(L,Mark)
4J	Medium Dark Spruce Poly	3378(F,M)
5H	Light Fawn Poly	3379(F,M)
5Q	Medium Dark Brown	2616(F,M)
5V	Dark Brown Poly	3399 (COugar XR7)
6B	Sand Poly	3287(F,M)
6J	Light Gold Poly	3391(Mark)
6N	Bright Yellow	3289(F,M)
6R	Cream	3380(Mark)
6V	Tan	2836(F,M)
7B	Light Spruce Poly	3382(F,M)
7C	Bright Lime Green Poly	3383(F,M)

7M	Dark Pine Poly	3290(F,M)
8D	Medium Bittersweet Poly	3293(F,M)
8E	Bittersweet Poly	3294(L,Mark,V)
8L	Medium Dark Nutmeg Poly	3392(L,Mark)
8N	Dark Cordovan Poly	2920(F,M)
9D	White	2684(All)
11	Light Pewter Poly	3516(F,M)
12	Light Dove Gray	3295(F,M,V)
15	Silver Poly	33402(L,Mark)
23	Dark Red Poly	3047(L,Mark,V)
24	Medium Red	3384(F,L,M)
27	Bright Red	3296(F,M)
54	Dark Champagne Poly	3297(L,Mark)
55	Medium Fawn Poly	3386(F,M)
56	Medium Fawn Poly	3298(L,Mark,V)
57	Light Fawn Poly	3299(L,Mark,V)
61	Medium Yellow	3488(F,M)
62	Antique Cream	3051(F,M)
86	Pastel Chamois	2999(F,M)
89	Fawn	3387(F,M)

F-Ford, L-Lincoln, M-Mercury, Mark-Mark V

1982 FORD

1A	Light Pewter Poly	3475(L,Mark)
1C	Black	9000/9300(All)
1G	Silver Poly	2593)F,M)
1H	Dark Pewter Poly	3476(F,M)
1L	Medium Dark Pewter Poly	3388(L,Mark)
1P	Medium Gray Poly	2967(F,M)
11	Light Pewter Poly	3516(F,M)
15	Silver Poly	33402(L,Mark)
17	Medium Pewter Poly	3385(F,M)
18	Light Pewter	3493(L,Mark)
2G	Bright Bittersweet	3278(F,M)
2H	Medium Red Poly	3168(F,M)
2K	Candy Apple Red	3279(F,M,)
2Q	Maroon Poly	2716(M)
23	Dark Red Poly	3047(L,Mark)
24	Medium Red	3384(F,M)
27	Brite Red	3296(F,M)
3D	Dark Blue Poly	3281(F,M)
3G	Light Blue Poly	3479(L,Mark)
3H	Medium Blue Poly	3033(F,M)
3K	Pastel Blue	3477(F,M)
3L	Dark Blue Poly	3171(F,M)
3M	Brite Blue Poly	3377(F,M)
3N	Brite Blue Poly	3494(F,M)
3P	Medium Blue Poly	3478(F,M)

3Q	Dark Blue Poly	3182(L,Mark)
34	Wedgewood Blue	3048(L,Mark)
4J	Medium Dark Spruce Poly	3378(F,M)
4T	Dark Teal Poly	3480(L)
41	Medium Light Teal Poly	3481(L)
5A	Medium Fawn Poly	3515(F,M)
5D	Dark Brown Poly	3482(L,Mark)
5H	Light Fawn Poly	3379(F,M)
5M	Medium Mulberry Poly	3483(L,Mark)
5V	Dark Brown Poly	3399(F,M)
5W	Medium Vaquero Poly	3174(F,M)
5Z	Light Fawn Poly	3484(L,Mark)
6B	Sand Poly	3287(F,M)
6V	French Vanilla Poly	3485(L,Mark)
6Y	Medium French Vanilla	3486(F,M)
6Z	Pastel French Vanilla	3487(All)
61	Medium Yellow	3488(F,M)
62	Antique Cream	3051(F,M)
68	French Vanilla Poly	3489(F,M)
69	Dark Curry Brown Poly	3490(F,M)
7A	Brite Lime Green Poly	3513(F,M)
7B	Light Spruce Poly	3382(F,M)
8G	Medium Bittersweet Poly	3514(F,M)
8N	Dark Cordovan Poly	2920(F,M)
8T	Medium Dark Mulberry Poly	3491(L,Mark)
87	Medium Fawn Poly	3492(L,Mark)
89	Fawn	3387(F,M)
9D	White	2684(All)
9G	Pastel Opal	90196(L)
67D	Revlon Scoundrel Purple	51128(M)

F-Ford, L-Lincoln, Mark-Mark V, M-Mercury

1983 FORD

1B	Medium Charcoal Poly	3562(F,M)
1C	Black	9000/9300(All)
1G	Silver Poly	2593)F,M)
1H	Dark Pewter Poly	3476(F,M)
1M	Light Charcoal	3563(F,M)
1Q	Silver Poly	3590(F,M,L)
1T	Light Pewter Poly	3376(F,M)
1V	Light Pewter Poly	3587(F,M)
1Z	Medium Charcoal Poly	3576(L)
11	Light Pewter Poly	3516(F,M)
17	Medium Pewter Poly	3385(F,M)
2G	Bright Bittersweet	3278(F,M)
2H	Medium Red Poly	3168(F,M)

2L	Maroon	3280(F,M)
2S	Medium Dark Red Poly	3564(F,M)
2U	Medium Red Poly	3577(F,M,L)
24	Medium Red	3384(F,M)
27	Brite Red	3296(F,M)
28	Dark Royal Blue Poly	3662(F)
3D	Dark Blue Poly	3281(F,M)
3F	Light Medium Blue	3170(F,M)
3H	Medium Blue Poly	3033(F,M)
3K	Pastel Blue	3477(F,M)
3L	Dark Blue Poly	3171(F,M)
35	Light Cadet Blue Poly	3611(F,M)
36	Midnight Blue	3566(F,M)
38	Medium Light Cadet Blue	3565(F,M)
4A	Dark Teal Poly	3567(F,M)
4W	Light Teal Poly	3574(F,M)
4Y	Loden Green	3568(F,M)
44	Dark Teal Poly	3578(L)
45	Light Teal Poly	3579(L)
5C	Dark Cadet Blue Poly	3588(F,M)
5M	Medium Mulberry Poly	3483(L,Mark)
5P	Medium Dark Mulberry Poly	3580(L)
5R	Dark Cadet Blue Poly	3581(F,M,L)
5U	Walnut Poly	3569(F,M)
5V	Dark Brown Poly	3399(F,M)
5W	Medium Vaquero Poly	3174(F,M)
6B	Sand Poly	3287(F,M)
6F	French Vanilla Poly	3582(F,M,L)
6K	Light French Vanilla Poly	3601(Mark)
6Y	French Vanilla	3486(F,M)
6Z	Pastel French Vanilla	3487(All)
61	Medium Yellow	3488(F,M)
62	Antique Cream	3051(F,M)
68	French Vanilla Poly	3489(F,M)
8G	Medium Bittersweet Poly	3514(F,M)
8N	Dark Cordovan Poly	2920(F,M)
8T	Medium Dark Mulberry Poly	3491(L,Mark)
87	Medium Fawn Poly	3492(L,Mark)
89	Fawn	3387(F,M)
9D	White	2684(All)
9J	Pastel Desert Sand Poly	3589(L)
9L	Oxford White	3620(F,M)
9N	Medium Desert Tan Poly	3575(F,M)
9P	Desert Tan	3570(All)
9Q	Light Desert Tan	3571(All)
9S	Walnut Poly	3583(F,M)
9T	Walnut	3572(F,M)
9Z	Midnight Cadet Blue Poly	3584(L)
91	Pastel Cadet Blue Poly	3585(F,M,L)
92	Dark Charcoal Poly	3586(F,M,L)

F-Ford, L-Lincoln, Mark-Mark VI, M-Mercury

1984 FORD

1B	Medium Charcoal Poly	3562(F,M)
1C	Black	9000(F,M)
1E	Silver Poly	3621(F,M)
1M	Light Charcoal	3563(T)
1Q	Silver Poly	3590(F,L,M)
1R	Black	9000(L)
1U	Lt. Oxford Gray	3690(L,M)
1Y	Dark Charcoal Poly	3647(C,T,L)
1Z	Medium Charcoal Poly	3576(F,L,M)
2A	Medium Canyon Red Poly	3644(F,M)
2C	Midnight Canyon Red Poly	3636(F,M)
2E	Light Canyon Red	3637(F,M)
2J	Midnight Canyon Red	3638(F,L,M)
2U	Medium Red Poly	3577(F,L,M)
2W	Midnight Wine	3849(L)
24	Medium Red	3384(F,M)
27	Brite Red	3296(F,M)
3A	Dark Blue Poly	3649(C,T,L)
3H	Medium Blue Poly	3033(F,M)
3L	Dark Blue Poly	3171(F,M)
35	Light Cadet Blue Poly	3611(F,M)
38	Medium Light Cadet Blue	3565(F,M)
4B	Light Sage Poly	3650(L)
4C	Medium Sage Poly	3651(L)
5C	Dark Cadet Blue Poly	3588(F,M)
5R	Dark Cadet Blue Poly	3581(F,L,M)
5U	Dark Walnut Poly	3568(F,M)
52	Medium Dark Cadet Blue Poly	3652(L)
53	Medium Desert Tan	3657(F,M)
6C	Light Wheat	3639(F,L,M)
6D	Wheat	3640(C,T)
6E	Wheat Poly	3653(L)
6G	Medium Dark Wheat Poly	3654(L)
6Z	Pastel French Vanilla	3487(F,M)
68	French Vanilla Poly	3489(F,M)
8Q	Light Desert Tan	3641(F,L,M)

8S	Pastel Desert Tan	3642(F,L,M)
9C	Brite Copper Poly	3646(F,M)
9J	Pastel Desert Tan Poly	3589(C,T)
9L	Oxford White	3620(F,L,M)
9N	Desert Tan Poly	3575(F,M)
9S	Walnut Poly	3583(C,L,T)
9W	Dark Charcoal Poly	3643(F,M)
91	Pastel Cadet Blue Poly	3585(L)

C-Cougar, F-Ford, M-Mercury,
L-Lincoln, T-Thunderbird

1985 FORD

1B	Medium Charcoal Poly	3562(F,M)
1C	Black	9000(F,M)
1E	Silver Poly	3621(F,M)
1M	LightCharcoal	3563(T)
1Q	Silver Poly	3590(F,L,M)
1R	Black	9000(L)
1U	Oxford Gray	3690(F)
1Y	Dark Charcoal Poly	3647(L)
1Z	Medium Charcoal Poly	3576(F,M)
2A	Mid. Canyon Red Poly	3644(F,M)
2C	Medium Canyon Red Poly	3636(F,M)
2E	Light Canyon Red	3637(C,T)
2J	Mid. Canyon Red	3638(F)
2M	Medium Canyon Red Poly	3717(F,L,M)
2R	Jalapena Red	3718(F)
2T	Dark Canyon Red Poly	3719(L)
3B	Mid. Regatta Blue Poly	3720(L)
3L	Dark Blue Poly	3171(F,M)
3M	Pastel Regatta Blue	3121(F,M)
3S	Pastel Regatta Blue Poly	3722(L,M)
3U	Mid. Regatta Blue	3723(F)
3V	Light Regatta Blue Poly	3724(F,M)
3Y	Medium Regatta Blue Poly	3725(F,L,M)
4B	Light Sage Poly	3650(L,M)
4C	Medium Sage Poly	3651(F,L)
4E	Dark Sage	3726(F)
4M	Dark Slate Poly	3728(F,M)
51	Medium Dark Fire Red	3656(F)
56	Medium Rosewood Poly	3729(F)
57	Dark Rosewood Poly	3730(L)
6C	Light Wheat	3639(L)
8H	Light Sandalwood Poly	25473(L)
8L	Sand Beige	3733(F,M)
8M	Sand Beige Poly	3734(F,L)
8U	Medium Sand Beige Poly	3735(L,M)
8W	Dark Sable Poly	3736(L)
8Y	Dark Sable Poly	3737(F,M)
8Z	Sandalwood Poly	3738(L)
81	Dark Sable	3739(F)

86	Medium Sand Beige	3740(F,M)
9C	Bright Copper Poly	3646(F)
9L	Oxford White	3620(F,L,M)

C-Cougar, F-Ford, L-Lincoln,
M-Mercury, T-Thunderbird

1986 FORD

1B	Med. Charcoal Poly	3562(F,M)
1C	Black	9000(F,L,M)
1D	Smoke Poly	3842(F,M)
1E	Silver Poly	621(F,M)
1G	Lt. Smoke Poly	3808(F,M)
1J	Graphite Poly	3844(L)
1K	Smoke	3845(F,L,M)
1L	Crystal Poly	3846(L)
1Q	Silver Poly	3590(F,L,M)
1R	Black	9700(C,L,T)
1V	Titanium Poly	3847(L)
14	Silver Poly	3811(F,M)
18	Smoke Poly	3848(F,M)
2A	Med. Canyon Red Poly	3644(F,M)
2C	Midnight Canyon Red Poly	3636(F,M)
2E	Lt. Canyon Red	3637(C,M,T)
2J	Midnight Canyon Red	3638(F,M)
2M	Med. Canyon Red Poly	3717(F,M)
2R	Jalapena Red	3718(F,M)
2T	Dk. Canyon Red Poly	3719(L)
2Y	Midnight Wine	3918(C,T)
28	Med. Canyon Red Poly	3850(M)
3J	Lt. Regatta Blue Poly	3851(F,M)
3L	Dk. Blue Poly	3171(F,M)
3R	Med. Shadow Blue Poly	3852(F,M)
3S	Pastel Regatta Blue Poly	3722(L)
3U	Midnight Regatta Blue	3723(F,M)
3V	Lt. Regatta Blue Poly	3724(F,M)
3W	Med. Regatta Blue Poly	3725(F,M)
3Y	Regatta Blue Poly	3725(F,L,M)
33	Lt. Regatta Blue Poly	3854(M)
37	Regatta Blue Poly	3855(F,M)
4E	Dark Sage	3726(F,M)
4F	Deep Aegean Poly	3856(L)
4G	Lt. Aegean Poly	3809(F,M)
4H	Med. Aegean Poly	3857(M)
4M	Dk. Slate Poly	3728(F,M)
46	Lt. Aegean Poly	3858(F,M)
48	Med. Aegean Poly	3859(F,M)
5A	Taupe	3860(F,M)
5B	Driftwood Poly	3861(M)
5H	Dk. Taupe Poly	3863(L)
5E	Lt. Taupe Poly	3862(C,L,T)
5H	Dk. Taupe Poly	3863(L)

55	Med. Taupe Poly	3864(F,M)
7A	Spinnaker Blue	3810(F,L,M)
7B	Shadow Blue Poly	3865(F,M)
7C	Deep Shadow Blue	3866(C,L,T)
8A	Med. Sand Beige	3867(F,M)
8D	Deep Sandalwood Poly	3868(L)
8L	Sand Beige	3733(F,L,M)
8M	Sand Beige Poly	3734(F,L,M)
8T	Med. Sandalwood Poly	3869(L)
8U	Med. Sand Beige	3735(L)
8Y	Dark Sable Poly	3737(F,M)
8Z	Sandlewood Poly	3870(F)
81	Dark Sable	3739(C,T)
86	Med. Sand Beige	3740(F,M)
9L	Oxford White	3620(F,L,M)

C-Cougar, L-Lincoln, F-Ford,
T-Thunderbird, M-Mercury

1987 FORD

1C	Black	9000(F,L,M)
1D	Smoke Poly	3842(F,M)
1E	Silver Poly	3621(F,M)
1J	Graphite Poly	3844(L)
1K	Smoke	3845(F,M)
1L	Crystal Poly	3846(L)
1Q	Silver Poly	3590(F,L,M)
1R	Black	9700(F,M)
1V	Titanium Poly	3847(L)
14	Silver Poly	3811(F,M)
18	Smoke Poly	3848(F,M)
2C	Mid. Canyon Red Poly	3636(F,M)
2D	Med. Scarlet	3935(F,M)
2H	Med. Cabernet	3936(F,M)
2N	Dk.Cabernet Poly	3937(L)
26	Brite Red Poly	3939(F,M)
3R	Med. Shadow Blue Poly	3852(F,M)
3S	Lt. Regatta Blue	3722(L)
3U	Mid. Regatta Blue	3723(F,L,M)
33	Lt. Regatta Blue	3854(F,M)
4L	Med. Prairie Moss Poly	3941(L)
48	Aegean Poly	3859(F,M)
5A	Taupe	3860(F,M)
5B	Driftwood Poly	3861(F,M)
5Q/5K	Deep Taupe Poly	3941(L)
5W	Lt. Taupe Poly	3944(C,T)
55	Med. Taupe Poly	3864(F,M)
6H	Jonquil	3945(L)
7A	Spinnaker Blue	3810(F,M)
7B	Shadow Blue Poly	3865(F,M)
7H	Brite Regatta Blue Poly	3745(Mustang)
7N	Dark Shadow Blue Poly	3946(F,M)

7P	Med. Shadow Blue Poly	3947(F,L,M)
7Q	Dark Shadow Blue Poly	3948(L)
77	Med. Shadow Blue Poly	3949(F,M)
8A	Med. Sand Beige	3867(F,L,M)
8D	Deep Sandalwood	3868(L)
8H	Lt. Sandalwood	3950(L)
8L	Sand Beige	3733(F,L,M)
8R	Lt. Sandalwood	3951(L)
8Y	Dk.Sable Poly	3737(F,M)
8Z	Sandalwood	3870(F,M)
87	Rose Quartz Poly	3952(L)
9L	Oxford White	3620(F,L,M)

C-Cougar, L-Lincoln, F-Ford,
M-Mercury, T-T-Bird

1988 FORD

1C	Black	9000/9300(F,L,M)
1D	Smoke Poly	3842(F,M)
1E	Silver Poly	3621(F,M)
1K	Smoke	3845(F,M)
1T	Titanium	4057(F,M)
11	Lt. Titanium	4121(L)
12	Ebony	9700(L)
19	Crystal	4058(L,M)
2D	Med. Scarlet	3935(F,T)
2G	Med. Cabernet Poly	4059(F,M)
2H	Med. Cabernet	3936(F,L,M)
2Y	Midnight Wine	3918(F,M)
26	Brite Red	3939(F,M)
3R	Med. Shadow Blue Poly	3852(F,M)
3U	Mid. Regatta Blue	3723(F)
31	Mid. Regatta Blue	4060(F,L,M)
44	Mid. Cabernet	4123(L)
48	Aegean	3859(F)
49	Graphite Poly	4076(F)
5L	Cinnabar	4061(F,L,M)
6B	Driftwood Poly	4062(F)
6K	Lt. Sandalwood	4124(F)
6L	Dk. Sable Poly	4063(F)
6Q	Deep Sandalwood Poly	4064(F,M)
6W	Med. Sandalwood	4065(F,M)
62	Med. Sandalwood	4066(F,L,M)
63	Lt. Sandalwood	4067(F,L,M)
66	Mimosa	4068(F)
7A	Spinnaker Blue	3810(F,M)
7F	Twilight Blue	4069(F,M)
7H	Bright Regatta Blue Poly	3745(F)
7K	Dk. Shadow Blue	3955(L,M)
7N	Dk.Shadow Blue Poly	3946(F,M)
77	Med. Shadow Blue	3009(F)
78	Shadow Blue	4070(F,L,M)

8L	Sand Beige	3733(F)		C2	Bright Red	72875(P)
8N	Rose Quartz	4071(F,L,M)		C7	Bright Blue Poly	16251(P)
8R	Lt. Sandalwood	3951(F,L,M)		C9	Oxford White	90373(P)
9L	Oxford White	3620(F,L,M)		D1	Silver Poly	34190(P)
9N	Med. Dark Titanium Poly	4125(F)		D2	Bright Red Poly	72881(P)
9P	Dk. Cabernet	4073(L,M)		E1	Dk. Titanium Poly	34113(P)
9R	Graphite Poly	3878(F)		E2	Garnet Poly	51243(P)
9Z	Silver Poly	4074(F,M)		F1	Crystal Clear Poly	34112(P)
91	Smoke Poly	4075(F,M)				
96	Med. Dk. Titanium Poly	4072(F)				

F-Ford, M-Mercury, L-Lincoln, P-Probe

F-Ford, M-Mercury, L-Lincoln

1989 FORD

1990 FORD

1C	Black Solid	9000/9300(F,L,M)		AA	Pastel Adobe	4170(F)
1D	Smoke Poly	3842(F)		AC	Med. Bisque Poly	4218(L)
1K	Smoke Solid	3845(F)		AD	Bisque Frost Poly	4206(L)
11	Lt. Titanium	4121(F,L,M)		AG	Race Yellow Solid	4207(F)
12	Ebony	9700(L,M)		AH	Pastel Alabaster	4208(F,L,M)
17	Titanium	4160(L,M)		AJ	Alabaster Solid	4209(F,M)
2D	Med. Scarlet	3935(F,M)		AK	Pastel Alabaster Solid	4219(L)
2F	Current Red	4161(F,M)		AP	Sandalwood Frost Poly	4210(F,L,M)
2H	Med. Cabernet Solid	3936(F,M)		AW	Med. Sandalwood Poly	4066(F,L,M)
2S	Bright Currant Red	4162(F,L,M)		AX	Lt. Sandalwood Poly	4067(F)
21	Vermilion	4163(F,M)		A2	Med. Sandalwood Poly	4245(P)
3K	Ultra Blue	4164(M)		A3	Pastel Adobe	4172(L)
30	Crystal Blue	4165(F,L,M)		A5	Lt. Sandalwood Poly	4243(P)
3R	Med. Shadow Blue Poly	3852(F)		CA	Med. Woodrose Poly	4211(F,M)
4N	Wild Strawberry Poly	4166(F,M)		CD	Woodrose Poly	4212(F,M)
4S	Dk. Titanium	4167(F,L,M)		EC	Currant Red Solid	4161(F,M)
49	Graphite	4076(F,M)		ED	Bright Currant Red	4162(F,L,M)
5C	Cinnabar	4061(F,L,M)		EE	Mid. Currant Red Poly	4173(F,L,M)
6F	Lt. Adobe	4168(F,M)		EG	Elec. Currant Red Poly	4213(F,L)
6V	Pastel Adobe	4170(F,M)		EH	Med. Cabernet Solid	3936(F)
62	Med. Sandalwood	4066(F,L,M)		EL	Wild Strawberry Poly	4166(F)
63	Lt. Sandalwood	4067(F,L,M)		EM	Med. Red Solid	3954(F)
66	Mimosa Solid	4068(F,M)		EP	Vermillion Solid	4163(F)
7A	Spinnaker Blue Solid	3810(F,M)		E2	Wild Strawberry Poly	4249(P)
7E	Lt. Crystal Blue	4171(F,M)		E3	Vermillion Red	4242(P)
7F	Twilight Blue	4069(F,L,M)		E4/6D	Vermillion Solid	4217(F,M)
7H	Brt. Regatta Blue Poly	3745(F)		KA	Crystal Blue Poly	4165(F,M)
7N	Dk. Shadow Blue Poly	3946(F)		KC	Lt. Crystal Blue Poly	4248(P)
8N	Rose Quartz	4071(F,L,M)		MA	Lt. Crystal Blue Poly	4171(F,M)
8S	Pastel Adobe	4172(L)		MD	Cl. Crystal Blue Frost Poly	4214(L)
9G	Mid Currant Red	4173(F,L,M)		ME	Med. Regatta Blue Poly	4060(L)
9L	Oxford White Solid	3620(F,L,M)		MH	Spinnaker Blue Solid	3810(F,M)
9Z	Silver	4074(F,M)		MK	Twilight Blue Poly	4069(F,L,M)
A8	Lt. Sandalwood Poly	25983(P)		MM	Ultra Blue Poly	4164(F)
B7	Dk. Shadow Blue Poly	16250(P)		MO	Mid. Regatta Blue	3723(F)
B8	Med. Sandalwood Poly	25984(P)		M1	Ultra Blue Poly	4250(P)
C1	Black	9000/9300(P)		M2	Twilight Blue Poly	4246(P)
				PA	Deep Jewel Green	4215(F)
				UA	Ebony	9700(L)

YC	Black Solid	9000/9300(F,M)
YD/I3	Pastel Titanium Solid	4261(F,M)
YF	Lt. Titanium Poly	4121(F,M)
YK	Med. Dk. Titanium Poly	4125(T)
YN	Silver Poly	4074(F,L)
YO	Oxford White Solid	3620(F,L,M)
YT	Graphite Poly	4076(F)
YU	Dk. Titanium Poly	4167(F,L,M)
YX	Titanivm Frost Poly	4216(L)
Y4	Brilliant Black	9000/9300(P)
Y5	Oxford White	4241(P)
Y6	Lt. Titanium Poly	4247(P)
Y7	Dk. Titanium Poly	4244(P)
Z6	Smoke	3845(F)
	Silver Poly	4262(F,L)

F-Ford, M-Mercury, L-Lincoln,
P-Probe, T-Tempo & Topaz

1991 FORD

AB	Sandalwood Spice Poly	4292(F,M,TR)
AC	Med. Bisque Poly	4218(L)
AD	Bisque Frost Poly	4206(L)
AG	Race Yellow Solid	4207(F)
AH	Pastel Alabaster	4208(F,M,TR)
AJ	Alabaster Solid	4209(F,M)
AK	Pastel Alabaster Solid	4219(F,M)
AL	Medium Alabaster	4285(F,M)
AP	Sandalwood Frost Poly	4210(F,M)
A2/A3	Med. Sandalwood Poly	4245(P)
A5	Lt. Sandalwood Poly	4243(P)
CA	Med. Woodrose Poly	4211(F,M)
CD	Woodrose Poly	4212(F,M)
DC	Med. Mocha Poly	4283(F,M,L)
DD	Mocha Frost Poly	4282(F,M,L)
EA	Cardinal Red	4325(C)
EB	Indigo Blue Poly	4326(C)
EC	Currant Red Solid	4161(F,M)
EE	Med. Currant Red Poly	4173(F,M)
EG	Elec. Currant Red Poly	4213(F,M,L)
EH	Med. Cabernet Solid	3936(F)
EL	Wild Strawberry Poly	4166(F,M,TR)
EM	Med. Red Solid	3954(F,M)
EP	Vermillion Solid	4163(F,TR)
ER	Dk. Cranberry Poly	4288(L)
EW	Lt. Cranberry Poly	4293(L)

EX	Med. Cranberry Poly	4287(L)
E2	Wild Strawberry Poly	4249(P)
E3	Vermillion Red	4242(P)
E4	Vermillion Solid	4217(F,M)
KA	Crystal Blue Poly	4165(F,M)
KB	Med. Amethyst Frost Poly	4286(F,M,L)
KC	Lt. Crystal Blue Poly	4248(P)
K2	Atlantic Blue Solid	4290(F,M)
MA	Lt. Crystal Blue Poly	4171(F,M,TR)
MB	Pastel St. Blue Frost Poly	4284(F,M)
MD	Clr. Crystal Blue Frost Poly	4214(F,M,L)
ME	Med. Regatta Blue Poly	4060(L)
MG	Brt. Regatta Blue Poly	3745(F,M)
MK	Twilight Blue Poly	4069(F,M,L,TR)
MM	Ultra Blue Poly	4164(F)
M1	Ultra Blue Poly	4250(P)
M2	Twilight Blue Poly	4246(P)
PA	Deep Jewel Green Poly	4215(F)
PC	Reef Blue Poly	4428(T)
PK	Chesapeke Blue Poly	4429(T)
UA	Ebony	9700(F,M,L)
W3	D. Charcoal Poly	4327(C)
YA	Polar White	4323(C)
YB	Platinum Silver Poly	4324(C)
YC	Black Solid	9000/9300(F,M,TR)
YD	Black	9700(F,M,L)
YD	Pastel Titanium Solid	4261(F,M)
YF	Lt.Titanium Poly	4121(F,M,TR)
YG	Med. Titanium Poly	4291(F,M,L,TR)
YK	Med. Dk. Titanium Poly	4125(F,M)
YO	Oxford White Solid	3620(F,M,TR)
YU	Dk. Titanium Poly	4167(F,M)
YX	Titanium Frost Poly	4216(F,M,L)
Y4	Brilliant Black	9000/9300(P)
Y5	Oxford White	4241(P)
Y6	Lt. Titanium Poly	4247(P)
Y7	Dk. Titanium Poly	4244(P)
YZ	White	4289(F,M)

F- Ford, M- Mercury, L- Lincoln,
P-Probe, TR- Tracer, T- Tempo &
Topaz, C-Capri

Chapter 9 GENERAL MOTORS, 1979-1995

PAINT COLOR CODE		PPG CODE
1979 GENERAL MOTORS		
10	Classic White	8631(Cor.)
11	White	2058(B,C,O,P,Cad.)
13	Silver Poly	2953(Cor.)
15	Silver Poly	3076(B,C,O,P,Cad.)
16	Gray Poly (Two-Tone)	3077(B,C,O,P)
19	Black	9300(B,C,O,P,Cad.,Cor.)
21	Pastel Blue	3119(B,C,O,P)
22	Light Blue Poly	2955(B,C,O,P,Cad.)
24	Bright Blue Poly	3120(B,C,O,P)
28	Frost Blue	3080(Cor.)
29	Dark Blue Poly	3121(B,C,O,P,Cad.)
33	Caramel Firemist Poly	3127(B,O)
35	Dark Red (Two-Tone)	72347(Cor.)
36	Medium Blue (Two-Tone)	15565(Cor.)
37	Medium Beige	24663(Cor.)
38	Beige Poly	24706(B)
39	Oyster White (Two-Tone)	3128(Cor.)
40	Pastel Green	3122(B,C,O,P)
41	Dark Aqua Poly	3130(Cad.)
44	Medium Green Poly	3123(B,C,O,P)
47	Indy Silver Poly	33323(Cor.)
49	Blackwatch Green	3083(Cad.)
50	Special Edition Gold Poly	3219(P)
51	Bright Yellow	3084(B,C,O,P)
52	Yellow	3072(Cor.)
54	Light Yellow	3085(B,C,O,P, Cad.)
55	Gold Poly (Two-Tone)	24430(O)
56	Burnished Gold Poly(Two-Tone)	3086(P)
58	Dark Green Poly	3140(Cor.)
59	Frost Beige	3087(Cor.)
61	Medium Beige	3124(B,C,O,P)
62	Pastel Beige	3131(Cad.)
63	Camel Poly	3125(B,C,O,P)
67	Dark Brown Poly	2656(Cor.)
68	Dark Gold Poly	3132(Cad.)
69	Dark Brown Poly	3126(B,C,O,P,Cad.)
72	Red	2973(Cor.)
75	Red	3095(B,C,O,P)
76	Dark Cedar Poly	3133(Cad.)
77	Carmine Poly	3096(B,C,O,P)
78	Carmine	3129(Cad.)
79	Dark Carmine Poly	3098(B,C,O,P)
80	Redbird Red	72326(P)
82	Brown Poly	2656(Cor.)
83	Dark Blue Poly	3074(Cor.)
84	Yellow Beige Poly(Two-Tone)	82432(B)
85	Medium Blue Poly (Two-Tone)	2980(B,C,O,P)
89	Light Gray	3134(Cad.)
90	Lt.Silver Gray Firemist Poly	3135(Cad.)
91	LightGold Firemist Poly	3136(Cad.)
92	Light Aqua Firemist Poly	3137(Cad.)
93	Light Cedar Firemist Poly	3138(Cad.)
94	Medium Green Firemist Poly	3101(Cad.)
95	Light Blue Firemist Poly	2981(Cad.)
96	Medium Saddle Firemist Poly	3103(Cad.)
98	Charcoal Firemist Poly	3141(B,O)
99	Saffron Firemist Poly	2971(B,O)

B-Buick, C-Chevrolet, Cad.-Cadillac, Cor.-Corvette, O-Oldsmobile, P-Pontiac

1979-80 HURST/OLDS SPECIAL

11	Body Color White	2058
55	Accent Gold Poly	24430
19	Body Color Black	9300
55	Accent Gold Poly	24430

1980 GENERAL MOTORS

10	White	8631(Cor.)
11	White	2058(Cad.,.B,C,O,P)
13	Silver Poly	2953(Cor.)
15	Silver	3076(Cad.,B,C,O,P)
16	Gray Poly (Two-Tone)	3077(B,C,O,P)

19	Black	9300(B,C,O,P,Cad.,Cor.)
20	Sterling Silver Poly	3308(P)
21	Light Blue Poly	3205(Cad.,B,C,O,P)
22	Medium Blue Poly	3206(Cad.,B,C,O,P)
24	Bright Blue Poly	3217(C,P)
28	Dark Blue Poly	3235(Cor.)
29	Dark Blue Poly	3207(Cad.,B,C,O,P)
36	Medium Beige Poly	3259(Cad.)
37	Accent Yellow	3224(C,P)
38	Medium Beige	3124(B)
40	Lime Green Poly	3218(C,P)
41	Pastel Green	3139(Cad.)
44	Dark Green Poly	3208(B,C,O,P)
47	Dark Brown Poly	2656(Cor.)
49	Blackwatch Green Poly	3083(Cad.)
50	Yellow	3209(B,C,O,P)
51	Bright Yellow	3219(C,P)
52	Yellow	3236(Cor.)
54	Light Yellow	3085(Cad.)
55	Gold Poly (Two-Tone)	24430(O)
56	Yellow	3225(C,P)
57	Gold Poly	3215(C,P)
58	Dark Green Poly	3140(Cor.)
59	Beige	3087(B,C,O,P,Cor.,Cad.)
61	Tan	3228(Cad.)
63	Light Camel Poly	3210(B,C,O,P)
67	Dark Brown Poly	3226(C,Cad.,P)
69	Medium Camel Poly	3211(B,C,O,P)
72	Red	2973(B,C,O,P)
75	Claret Poly	3220(B,C,O,P)
76	Dark Claret Poly	3212(Cor.,Cad.,B,C,O,P)
77	Cinnabar	3213(B,C,O,P)
78	Carmine	3129(Cad.)
79	Red Orange	3221(C,P)
80	Rust Poly	3222(C,P)
83	Red	3237(Cor.)
84	Charcoal Poly	3223(C,P)
85	Vapor Gray	3214(B,C,O,P)
89	Light Gray	3134(Cad.)
90	Light Blue Firemist Poly	3229(Cad.)
91	Tan Firemist Poly	3230(Cad.)
92	Claret Firemist Poly	3231(Cad.)
94	Med. Silver Gray Firemist Poly	3232(Cad.)
96	Saddle Firemist	3103(Cad.)

	Poly	
97	Medium Brown Firemist Poly	3222(B,O)
98	Charcoal Firemist Poly	3141(B,O)
99	Dark Brown Firemist Poly	3234(B,O)

B-Buick, C-Chevrolet, Cad.-Cadillac.
Cor.-Corvette, O-Oldsmobile, P-Pontiac

1981 GENERAL MOTORS

06	Mahogany Poly	2864(Cor.)
08	Gray Poly	3077(C,B,O,P)
09	Red	3095(C,P)
10	White	8631(Cor.)
11	White	2058(Cad.,C,B,O,P)
13	Silver Poly	2953(Cor.)
16	Silver Poly	3308(Cad.,C,B,O,P)
19	Black	9300(B,C,O,P,Cad.,Cor.)
20	Bright Blue Poly	3309(C,P)
21	Light Blue Poly	3310(Cad.,C,B,O,P)
22	Medium Blue Poly (Two-Tone)	3311(C,B,O,P)
24	Bright Blue Poly	3340(Cor.)
28	Dark Blue Poly	3074(Cor.)
29	Dark Blue Poly	3207(Cad.,C,B,O,P)
31	Medium Aqua Poly	3312(Cad.)
32	Pepper Green Poly	3313(Cad.)
33	Silver Poly	3402(Cor.)
34	Light Camel Poly	3210(B,O)
35	Pastel Waxberry	3314(Cad.,C,B,O,P)
36	Light Waxberry Poly	3315(Cad.,C,B,O,P)
37	Medium Waxberry Poly (2-Tone)	3316(C,B,O,P)
38	Dark Blue Poly	3473(Cor.)
39	Charcoal Poly	3414(Cor.)
40	Yellow	3225
44	Pastel Jadestone	3317(Cad.
45	Light Jadestone Poly	3318(C,B,O,P)
47	Dark Jadestone Poly	3319(C,B,O,P)
48	Dark Green Poly	3320(C,B,O,P)
49	Dark Jadestone Firemist Poly	3321(B,O)
50	Beige	3474(Cor.)
51	Bright Yellow	3219(C,B,O,P)
52	Yellow	3236(Cor.)
54	Gold Poly	3322(C,P)
56	Yellow	3324(C,P)

57	Citrus Orange Poly	3325(C,P)
58	Orange Poly	3326(C,B,O,P)
59	Beige	3087(C,B,O)
61	Tan	3228(Cad.)
63	Pastel Sandstone	3327(C,B,O,P)
67	Dark Brown Poly	3328(C,P)
68	Medium Sandstone Poly	3329(C,B,O,P)
69	Dark Sandstone Poly	3330(Cad.,C,B,O,P)
72	Light Maple Poly	3331(Cad.,C,B,O,P)
74	Dark Bronze Poly	3446(Cor.)
75	Spectra Red	3332(Cor.,C,B,O,P)
76	Dark Claret Poly	3212(Cad.)
77	Dark Maple Poly	3333(C,B,O,P)
78	Carmine	3129(Cad.)
79	Maroon Poly	3341(Cor.)
80	Autumn Red Poly	3445(Cor.)
83	Silver Fawn Firemist Poly	3334(B,O)
84	Charcoal Poly	3223(C,Cor.,P)
85	Medium Slate Firemist Poly	3335(B,O)
89	Light Gray	3134(Gray)
90	Light Blue Firemist Poly	3229(Cad.)
91	Tan Firemist Poly	3230(Cad.)
92	Claret Firemist Poly	3231(Cad.)
94	Medium Silver Gray Firemist Poly	3232(Cad.)
96	Medium Jadestone Firemist Poly	3336(Cad.)
97	Doeskin Firemist Poly	3337(Cad.)
98	Dark Claret Poly	3452(Cor.)
99	Mulberry Gray Firemist Poly	3338(Cad.)
19	Black	9300(B,C,O,P,Cad.,Cor.)

B-Buick, C-Chevrolet, Cad.-Cadillac,
Cor.-Corvette, O-Oldsmobile, P-Pontiac

1982 GENERAL MOTORS

10	White	8631(Cor.)
11	White	2058(Cad.,C,B,O,P)
13	Silver Poly	3402(Cor.)
16	Silver Poly	3308(Cad.,C,B,O,P)
E16	Silver Poly	3507(C,B,O,P)
19	Black	9300(B,C,O,P,Cad.,Cor.)
21	Light Blue Poly	3310(Cad.,C,B,O,P)
E21	Light Blue Poly	3506(C,B,O,P)
EN21	Light Blue Poly	15673(C,B,O,P)
22	Bright Blue Poly	3407(C,B,O,P)
24	Silver Blue Poly	3404(Cor.)
26	Dark Blue Poly	3410(Cor.)
29	Dark Blue Poly	3207(Cad.,C,B,O,P)
30	Medium Blue Poly	3412(Cad.)
31	Bright Blue Poly	3413(Cor.)
39	Charcoal Poly	3414(Cor.)
40	Silver Green Poly	3415(Cor.)
45	Light Jadestone Poly	3318(C,B,O,P)
E45	Light Jadestone Poly	3508(C,B,O,P)
EN45	Light Jadestone Poly	45741(C,B,O,P)
49	Dark Jadestone Poly	3421(C,B,O,P)
54	Yellow	3085(Cad.)
55	Gold Wing Poly	3424(C,B,O,P)
E55	Gold Wing Poly	3509(C,P)
56	Gold Poly	3426(Cor.)
57	Marigold	3427(C,B,O,P)
59	Silver Beige Poly	3429(Cor.)
61	Almond	3430(Cad.)
63	Pastel Sandstone	3327(C,B,O,P)
64	Dark Sable Poly	3431(Cad.)
67	Dark Gold Wing Poly	3439(C,B,O,P)
68	Medium Sandstone Poly	3329(C,B,O,P)
EG8	Medium Sandstone	3510(C,B,O,P)
70	Spectra Red	3512(Cor.)
72	Light Redwood Poly	3435(Cad.,C,B,O,P)
E72	Light Redwood Poly	3511(C,B,O,P)
74	Autumn Maple Firemist	3437(B.O)
75	Spectra Red	3332(C,B,O,P)
77	Dark Redwood Poly	3439(Cad.,C,B,O,P)
78	Dark Claret Poly	3440(C,B,O,P)
80	Slate Gray	3442(Cad.,C,B,O,P)
81	Carmine Poly	3443(Cad.)
82	Dark Brown Poly	3444(Cad.)
84	Charcoal Poly	3223(C,B,O,P)
85	Medium Slate Firemist	3335(B,O)
88	Dark Brown Firemist	3517(B,O)
90	Azure Blue Firemist	3229(Cad.)
91	Medium Spruce Firemist	3448(Cad.)
94	Medium Gray Firemist	3232(Cad.)
96	Light Sable Firemist	3449(Cad.)
97	Light Redwood Firemist	3450(Cad.)
98	Almond Firemist	3451(Cad.)
99	Dark Claret Poly	3452(Cor.)

1983 GENERAL MOTORS

10	White	8631(Cor.)
11	White	2058(Cad.,C,B,O,P)
15	Silver Sand Gray Poly	3520(Cad.,C,B,O,P)
15E	Silver Sand Gray Poly	3521(C,P)
15W	Silver Sand Gray Poly	3524(C,P)
16	Bright Silver Poly	3602(Cor.)
18	Medium Gray Poly	3603(Cor.)
19	Black	9300(B,C,O,P,Cad.,Cor.)
20	Light Blue Poly	3604(Cor.)
22	Light Royal Blue Poly	3522(Cad.,C,B,O,P)
23	Medium Blue Poly	3605(Cor.)
22E	Light Royal Blue Poly	3523(C,P)
22W	Light Royal Blue Poly	3527(C,P)
27	Med. Dark Royal Blue Poly	3323(Cad.,C,B,O,P)
27E	Medium Dark Royal Blue Poly	3545(C,P)
27W	Medium Dark Royal Blue Poly	3546(C,P)
39	Medium Beige	25360(O)
42	Light Grayfern Poly	3526(Cad.,C,B,O,P)
48	Dark Grayfern Poly	3528(Cad.,C,B,O,P)
50	Light Flax	3529(Cad.)
53	Gold Poly	3609(Cor.)
59	Cream Beige	3087(C,B,O,P)
60	Light Sand Gray	3530(Cad.,C,B,O,P)
62	Light Briar Brown Poly	3531(Cad.,C,B,O,P)
62E	Light Briar Brown Poly	3549(C,P)
62W	Light Briar Brown Poly	3550(C,P)
63	Light Bronze Poly	3606(Cor.)
64	Medium Beech Poly	3533(Cad.)
65E	Dark Gold Wing Poly	3615(C,P)
65W	Dark Cold Wing Poly	3616(C,P)
66	Dark Bronze Poly	3607(Cor.)
67	Dark Briar Brown Poly	3534(Cad.,C,B,O,P)
67E	Dark Briar Brown Poly	3551(C,P)
67W	Dark Briar Brown Poly	3552(C,P)
70	Spectra Red	3512(Cor.)
73	Light Maple Poly	3331(C,B,O,P)
74	Autumn Maple Firemist	3532(B,O,Cad.)
75	Spectra Red	3332(Cad.,C,B,O,P)
78	Dark Autumn Maple Poly	3553(Cad.,C,B,O,P)
81	Carmine Poly	3544(Cad.)
82E	Midnight Sand Gray Poly	3536(C,P)
82W	Midnight Sand Gray Poly	3537(C,P)
87	Medium Sand Gray Poly	3538(B,O)
90	Deep Royal Blue Firemist	3539(Cad.)
91	Light Flax Firemist	3540(Cad.,B,O)
92	Medium Briar Brown Firemist	3541(Cad.)
93	Light Beech Firemist	3542(Cad.)
94	Dark Beech Firemist	3543(Cad.)

1984 GENERAL MOTORS

06	Charcoal Poly	3612(B,P)
11	White	2058(Cad.,B,C,O,P)
13	Platinum	3661(Cad.)
15E	Silver Sand Gray Poly	3521(C,P)
15W	Silver Sand Gray Poly	3524(C,P)
16	Brite Silver Poly	3602(Cor.)
17	Silver Poly	3659(Cad.)
18	Medium Gray Poly	3603(Cor.)
19	Black	9300(B,C,O,P,Cad.,Cor.)
20	Light Blue Poly	3604(Cor.)
21	Balboa Blue	3660(Cad.)
22	Light Royol Blue Poly	3522(Cad.,B,C,O,P)
22E	Light Royal Blue Poly	3523(C,P)
22W	Light Royal Blue Poly	3527(C,P)
23	Medium Blue Poly	3605(Cor.)
25	Med. Royal Blue Poly	3612(P)
27	Med. Dark Royal Blue Poly	3525(Cad.,B,C,O,P)

27E	Med. Dark Royal Blue Poly	3545(C,P)
27W	Med. Dark Royal Blue Poly	3546(C,P)
28	Diplomat Blue	3662(Cad.)
33C	Bright Red	3681(Cor.)
36	Bright Blue	3692(P)
37C	Cameo Ivory	3668(Cad.)
39C	Sandalwood	3663(Cad.)
40C	White	3680(Cor.)
41C	Black	9300(Cor.)
42	Light Grayfern Poly	3526(Cad.,B,C,O,P)
43	Med. Grayfern Poly	3614(P)
44	Grayfern	3664(Cad.)
48	Dark Grayfern Poly	3528(Cad.,B,C,O,P)
50	Light Flax	3529(Cad.)
53	Gold Poly	3609(Cor.)
55	Gold Wing Poly	3424(P)
55A	D'Oro Gold	25430(Cad.)
59	Cream Beige	3087(B,C,O,P)
60	Light Sand Gray Poly	3530(Cad.,B,C,O,P)
62	Light Briar Brown Poly	3531(Cad.,B,C,O,P)
62E	Light Briar Brown Poly	3549(C,P)
62W	Light Briar Brown Poly	3550(C,P)
63	Light Bronze Poly	3606(Cor.)
64	Medium Beech Poly	3533(Cad.)
65E	Dark Gold Wing Poly	3615(C,P)
65W	Dark Gold Wing Poly	3616(C,P)
66	Bark Bronze Poly	3607(Cor.)
67	Dark Briar Brown Poly	3534(B,C,O,P)
67E	Dark Briar Brown Poly	3551(C,P)
67W	Dark Briar Brown Poly	3552(C,P)
68	Canyon Brown	3665(Cad.)
71	Red	3679(P)
72	Coronation Red	3666(Cad.)
73	Light Maple Poly	3331(Cad.,B,C,O,P)
74	Autumn Maple Firemist	3532(Cad.,B,O)
75	Spectra Red	3332(B,C,O,P)
78	Dark Autumn Maple Poly	3553(Cad.,B,C,O,P)
82E	Midnight Sand Gray Poly	3536(C,P)
82W	Midnight Sand Gray Poly	3537(C,P)
85	Saddle Tan Firemist	3682(B,O)
87	Med. Sand Gray Firemist	3538(B,O)
90	Deep Royal Blue Firemist	3539(Cad.)
92	Med. Briar Brown Firemist	3541(Cad.)
93	Light Beech Firemist	3542(Cad.)
94	Dark Beech Firemist	3543(Cad.)
95	Brittany Blue Firemist	3669(Cad.)
96	Desert Frost Firemist	3670(Cad.)
97	Frost Green Firemist	3671(Cad.)
98	Madiera Plum Firemist	3672(Cad.)
99	Black Cherry	3673(Cad.)

B-Buick, C-Chevrolet, Cad-Cadillac,
Cor-Corvette, O-Oldsmobile, P-Pontiac

1985 GENERAL MOTORS

11	White	2058(B,C,O,P,Cad.)
12	Silver Poly	3746(B,C,O,P,Cad.)
12E	Silver Poly	3747(C,P)
12W	Silver Poly	3748(C,P)
13	Silver Poly	3661(B,O,P,Cad.Cor.)
14	Light Gray Poly	3706(P)
15	Med. Gray Poly	3749(B,C,O,P,Cad.)
15E	Med. Gray Poly	3750(C,P)
15W	Med. Gray Poly	3751(C,P)
16	Silver Poly	3308(C,P)
17	Black Poly	3752(B,O,P)
18	Medium Gray Poly	3603(Cor.)
19	Black	9300(B,C,O,P,Cad.)
20	Light Blue Poly	3604(Cor.)
22	Lt. Blue Poly	3753(B,O,P,Cad.)
23	Medium Blue Poly	3605(Cor.)
25	Lt. Blue Poly	3753(B,C,O,P,Cad.)
26	Dark Blue	3755(C,P)
27	Med. Blue Poly	3756(B,O,P)
29	Dark Blue Poly	3737(B,O,Cad.)
30E	Bright Blue Poly	3758(C,P)
30W	Bright Blue Poly	3759(C,P)
31	Dk.Blue Poly	3760(B,C,O,P,Cad.)
37	Lt. Flax	3668(Cad.)
39	Sandstone	3663(B,O,Cad.)
40	White	3680(B,O,P,Cad.,Cor.)
41	Black	9300(B,O,P,Cad.,Cor.)
42	Light Sage Poly	3761(B,O,P)
43	Light Sage Poly	3762(B,C,O,P,Cad.)
46	Med. Sage Poly	3763(B,C,O,P,Cad.)
47	Med. Sage Poly	3764(B,O,Cad.)

49	Dk. Sage Poly	3765(B,O,P)
50	Yellow Gold	3766(C,P)
51	Yellow Gold	3767(C,O,P)
52	Sedona Tan	3768(B,O,P)
53	Gold Poly	3609(Cor.)
54	Yellow Beige	3769(Cad.)
56	Sandstone	3770(B,O,Cad.)
57	Dk.Sandstone Poly	3771(B,O,Cad.)
58	Lt. Chestnut Poly	3772(B,O,Cad.)
59	Cream Beige	3087(B,C,O,P,Cad.)
60E	Lt. Chestnut Poly	3773(C,P)
60W	Lt. Chestnut Poly	3774(C,P)
61	Dk. Chestnut Poly	3775(B,O,P)
62	Dk. Chestnut Poly	3776(B,C,O,P)
63	Light Bronze Poly	3606(Cor.)
64	Dk. Chestnut Poly	3777(B,O,P)
65	Lt. Russet Poly	3778(B,C,O,P,Cad.)
66	Dark Bronze Poly	3607(Cor.)
68	Dk. Sand Stone Poly	3665(B,O,Cad.)
69E	Russet Poly	3779(C,P)
69W	Russet Poly	3780(C,P)
70	Carmine	3781(B,C,O,P)
71	Red	3679(P)
72	Crimson Maple Poly	3666(B,O,Cad.)
73	Light Maple Poly	3331(Cad.)
74	Autumn Maple Firemist	3532(B,O,Cad.)
75	Blaze Red	3782(C,P)
76	Red Poly	3783(B,O,P)
77	Dk. Russet Poly	3784(Cad.)
78	Dk. Red	3785(C,P)
79	Dk. Red Poly	3786(B,C,O,P,Cad.)
81	Flame Red	3794(Cor.)
82	Med. Sand Gray Poly	3536(C,P)
84	Gun Metal Poly	3667(B,O,P,Cad.)
85	Saddle Tan Firemist	3682(B,O,Cad.)
86	Saddle Tan Firemist	3787(Cad.)
88	Dk. Gray Poly	3788(Cad.)
89	Very Dk. Sandstone Poly	3683(B,O)
90	Med. Blue Firemist	3789(Cad.)
91	Heather Firemist	3684(Cad.)
92	Canyon Red Firemist	3790(Cad.)
93	Heather Poly	3791(Cad.)
94	Med. Blue Firemist	3792(B,O,Cad.)
96	Lt. Sandstone Firemist	3670(B,O,Cad.)
97	Dk. Gray Firemist	3793(B,O,Cad.)
98	Dk. Crimson Maple Poly	3672(Cad.)
99	Black Cherry	3673(Cad.)

B-Buick, C-Chevrolet, O-Oldsmobile,
P-Pontiac, Cad.-Cadillac, Cor.-Corvette

1985 CADILLAC SALES NAMES

11	Cotillion White	2058
12	Silver Frost	3746
19	Platinum	3661
15	Academy Gray	3749
19	Sable Black	9300
22	Gossamer Blue	3753
25	Gossamer Blue	3754
29	Commodore Blue	3757
31	Commodore Blue	3760
37	Cameo Ivory	3668
39	Sandalwood	3663
40	Cotillion White	3680
41	Sable Black	9300
43	Aspian Green	3762
46	Sage Green	3763
47	Sage Green	3764
54	Chamois	3769
56	Sandalwood	3770
57	Burlwood Brown	3771
58	Laredo Tan	3772
59	Cream Beige	3087
68	Canyon Brown	3665
72	Coronation Red	3666
73	Burgundy	3331
74	Autumn Maple	3532
77	Russet	3784
79	Bordeaux Red	3786
84	Academy Gray	3667
85	Flaxen Firemist	3682
86	Desert Frost Firemist	3787
88	Deaville Gray	3788
90	Corinthian Blue	3789
91	Heather Firemist	3684
92	Cranberry Firemist	3790
93	Heather	3791
94	Corinthian Blue Firemist	3792
96	Desert Frost Firemist	3670
97	Charcoal Firemist	3793
98	Madiera Plum Pearlmist	3672
99	Black Cherry Pearlmist	3673

1986 GENERAL MOTORS

11	White	2058(B,C,O,P,Cad.)
12	Silver Poly	3746(B,C,O,P,Cad.)

13S	Silver Poly	3822(B,O,P,Cad.)		78	Dk. Carmine Poly	3901(B,O)
15	Med.Gray Poly	3749(B,C,O,P,Cad.)		79	Dk.Red Poly	3786(B,C,O,P,Cad.)
16	Silver Poly	3880(P)		80	Sandstone	3770(Cad.)
17	Black Poly	3752(B,C,O,P)		81	Bright Red	3794(P,Cam,Cor,F'Bd)
18	Med. Gray Poly	3603(Cor.)		82	Dark Gray Firemist	3793(Cad.)
19	Black	9300(B,C,O,P,Cad.)		83	Canyon Red Firemist	3790(Cad.)
20	Nassau Blue Poly	3881(Cor.)		84	Gunmetal Poly	3667(B,C,O,P,Cad.)
22	Lt. Blue Poly	3753(B,C,O,P,Cad.)		85	Med. Blue Firemist	3789(Cad.)
23	Bright Blue Poly	3882(Cam,F'Bd)		86	Med. Quartz Poly	3902(Cad.)
24	Bright Blue Poly	3883(B,C,O,P)		87	Lt.Sandstone Firemist	3787(Cad.)
25	Lt. Blue Poly	3754(B,C,O,P,Cad.)		88	Dark Gray Poly	3788(Cad.)
26	Bright Blue Poly	3884(P)		89	Autumn Maple Firemist	3532(Cad.)
27	Med. Blue Poly	3756(B,C,O,P)		90	White Diamond	3903(Cad.)
28	Black Sapphire Poly	3885(Cam,F'Bd)		91	Heather Firemist	3884(Cad.)
29	Dark Blue Poly	3757(B,C,O,P,Cad.)		92	Med. Emerald Firemist	3904(B,O,Cad.)
31	Dk. Blue Poly	3760(B,C,O,P,Cad.)		93	Heather Poly	3791(Cad.)
35	Yellow	3886(Cor.)		94	Med. Blue Firemist	3792(B,O,Cad.)
38	Sungold	3887(Cad.)		95	Sungold Firemist	3905(Cad.)
40	White	3680(B,C,O,P,Cad.,Cor.)		96	Dk. Driftwood Firemist	3906(B,O,Cad.)
41	Black	9700(B,C,O,P,Cad.,Cor,)		97	Lt. Quartz Firemist	3907(Cad.)
42	Lt. Sage Poly	3761(B,C,O,P)		98	Chamois Firemist	3908(Cad.)
43	Lt. Sage Poly	3762(B,C,O,P,Cad.)		99	Black Cherry Poly	3673(B,O,Cad.)
46	Med. Sage Poly	3763(Cad.)				
48	Black Emerald Poly	3888(B,O,Cad.)				
49	Dark Sage Poly	3765(B,C,O,P)				

B-Buick, C-Chevrolet, O-Oldsmobile, P-Pontiac, Cad.-Cadillac, Cor.-Corvette, Cam-Camaro, F'Bd-Firebird

51	Yellow Gold	3889(Cam.F'Bd.)
52	Sedona Tan	3768(B,C,O,P)
53	Gold Poly	3609(Cor.)
54	Yellow Beige	3769(B,C,O,P,Cad.)
55	Chamois	3890(Cad.)
56	Champagne Poly	3891(P)
57	Dk. Sandstone Poly	3771(Cad.)
58	Lt. Chestnut Poly	3772(B,C,O,P,Cad.)
59	Silver Beige Poly	3892
60	Champagne Gold Poly	3893(P,Cam,F'Bd)
61	Lt. Chestnut Poly	3775(B,C,O,P)
62	Dk. Chestnut Poly	3776(B,C,O,P)
63	Lt. Driftwood	3894(B,O,Cad.)
64	Dk. Chestnut Poly	3777(B,C,O,P)
65	Med. Driftwood Poly	3895(B,O,Cad)
66	Russet Poly	3896(Cam,F'bd,Cor.)
68	Midnight Russet Poly	3897(Cam,F'bd)
69	Dk. Beige Poly	3898
70	Carmine	3781(B,C,O,P)
71	Med. Red Poly	3899(B,C,O,P)
73	Lt. Maple Poly	3331(Cad.)
74	Flame Red Poly	3823(B,C,O,P,Cad.)
75	Garnet Red Poly	3900(B,O,Cad.)
77	Dk.Russet Poly	3784(Cad.)

1986 CADILLAC SALES NAMES

11	Cotillion White	2058
12	Silver Frost	3746
13S	Platinum	3822
15	Academy Gray	3749
19	Sable Black	9300
22	Gossamer Blue	3753
25	Gossamer Blue	3754
29	Commodore Blue	3757
31	Commodore Blue	3760
38	Sunburst Yellow	3887
40	Cotillion White	3680
41	Sable Black	9700
43	Aspen Green	3762
46	Sage Green	3763
48	Black Emerald	3888
54	Chamois	3769
55	Chamois	3769
57	Burlwood Brown	3771
58	Laredo Tan	3772
63	Driftwood	3894
65	Desert Beige	3895

73	Burgundy	3331	25	Lt. Blue Poly	3754(B,C,O,P,Cad.)
74	Crimson	3823	25	Lt. Blue Poly	3922(B,C,(Ok.City)
75	Garnet	3900	25	Lt. Blue Poly	3965(C,O,P,(Oshawa)
77	Russet	3784	27	Med.Sapphire Blue Poly	3967(B,C,O,P,Cad.)
79	Bordeaux Red	3786	28	Black Sapphire Poly	3885(B,C,O,P,Cad.)
80	Sandalwood	3770	29	Dk. Saphire Blue Poly	3968(B,C,O,P,Cad.)
82	Charcoal Firemist	3793	31	Dk. Blue Poly	3760(B,C,O,P,Cad.)
83	Cranberry Firemist	3790	31	Dk. Blue Poly	3723(B,C,(Ok.City)
84	Academy Gray	3667	31	Dk. Blue Poly	3969(C,O,P(Oshawa)
85	Corinthian Blue	3789	33	Lt. Driftwood Poly	3970(B,C,O,P,Cad.)
86	Quartz	3902	35	Yellow	3886(Cor.)
87	Desert Frost Firemist	3787	37	Lt.Silver Poly	3971(C,O,P,(Oshawa)
88	Deaville Gray	3788	38	Sungold	3887(B,O,Cad.)
89	Autumn Maple Firemist	3532	39	Lt.Saddle Poly	3972(C,O,P(Oshawa)
91	Heather Firemist	3684	40	White	3680(B,C,O,P,Cad.)
92	Emerald Firemist	3904	41	Black	9700(B,C,O,P,Cad.)
93	Heather	3791	42	Lt. Emerald Poly	3973(B,C,P,)
94	Corinthian Blue Firemist	3792	43	Lt. Sage Poly	3762(B,C,O,P)
95	Sungold Firemist	3905	43	Lt. Sage Poly	3924(B,C,(Ok City)
96	Dk. Driftwood Firemist	3906	43	Lt.Sage Poly	3974(C,O,P(Oshawa)
97	Quartz Firemist	3907	48	Black Emerald	3888(B,O,P,Cad.)
98	Almond Gold Firemist	3908	49	Dk. Saxony Green Poly	3975(Cad.)
99	Black Cherry Pearlmist	3673	51	Yellow Gold	3889(Cam,F'Bd)

1987 GENERAL MOTORS

11	White	2058(B,C,O,P,Cad.)	52	Copper Beige	3976(B,O,P)
11	White	3919(B,C,(Ok City)	53	Gold Poly	3609(Cor.)
11	White	3959(C,O,P(Oshawa)	54	Yellow Beige 3	769(B,C,O,P,Cad.)
12	Silver Poly	3746(B,C,O,P,Cad.)	54	Yellow Beige	3925(B,C,OkCity)
12	Silver Poly	3920(B.C.(Ok City)	54	Yellow Beige	3977(C,O,P(Oshawa)
12	Silver Poly	3960(C,O,P(Oshawa)	56	Gold Poly	3891(P)
13	Silver Poly	3661(Cor.)	57	Very Dk. Chestnut F.M.	3978(Cad.)
13S	Silver Poly	3822(B,C,O,P,Cad.)	58	Lt. Chestnut Poly	3772(B,C,O,P,Cad.)
14	Black Poly	3961(B,C,O,P,Cad.)	58	Lt. Chestnut Poly	3926(B,C,(Ok City)
15	Med. Gray Poly	3749(B,C,O,P,Cad.)	58	Lt. Chestnut Poly	3979(C,O,P.(Oshawa)
15	Med. Gray Poly	3921(B,C,(Ok City)	59	Silver Beige Poly	3892(Cor.)
15	Med. Gray Poly	3962(C,O,P(Oshawa)	62	Dk. Chestnut Poly	3776(B,C,O,P,Cad.)
16	Silver Poly	3880(P)	62	Dk. Chestnut Poly	3927(B,C(OkCity)
17	Pearl Gray	3963(Cad.)	62	Dk. Chestnut Poly	3981(C,O,P(Oshawa)
18	Med. Gray Poly	3603(Cor.)	63	Lt. Driftwood	3894(B,C,O,P,Cad.)
19	Black	9300(B,C,O,P,Cad.)	64	Med. Copper Poly	3982(B,O,P)
19	Black	9700(B,C,(Ok City)	66	Dk. Orange Poly	4008(Cor.)
19	Black	9700(C,O,P(Oshawa)	67	Lt.Copper Poly	3984(B,O,P)
21	Bright Blue Poly	3884(B,C,O,P)	68	Midnight Russet Poly	3897(Cam.F'Bd)
22	Lt.Sapphire Blue Poly	3964(B,C,O,P,Cad.)	69	Dk. Beige Poly	3898(Cor.)
23	Bright Blue Poly	3882(Ca.,F'Bd)	70	Carmine	3781(B,C,O,P,Cad)
24	Bright Blue Poly	3883(B,C,O,P,Cad.)	71	Med. Red Poly	3899(B,C,O,P)
			71	Med. Red Poly	3928(B,C,(OkCity)

71	Med.RedPoly	3985(C,O,P(Oshawa)
72	Med. Garnet Red Poly	3986(B,C,O,P,Cad.)
74	Flame Red Poly	3823(Cam,Cor,F'Bd)
75	Med.Garnet Red Poly	3986(Cad.)
76	Dk.Garnet Red Poly	3989(B,O,P,Cad.)
77	Med. Red Poly	3990(P)
78	Med. Rosewood Poly	3991(B,C,O,P,Cad.)
79	Dk. Red Poly	3786(B,C,O,P,Cad.)
79	Dk. Red Poly	3929(B,C,(OkCity)
79	Dk. Red Poly	3992(C,O,P(Oshawa)
80	Lt. Rosewood Poly	3993(B,C,O,P,Cad.)
81	Bright Red	3794(Cam,Cor.F'bd,P)
82	Med.Rosewood Poly	3994(B,C,O,P,Cad.)
83	Med. Sapphire Blue F.M.	3995(B,O,Cad.)
84	Gunmetal Poly	3667(B,C,O,P,Cad.)
85	Med. Blue Firemist	3789(Cad.)
86	Dk. Rosewood Poly	3996(B,C,O,P,Cad.)
87	Gunmetal Poly	4009(Cam,F'Bd)
88	Dark Gray Poly	3788(Cad.)
89	Autumn Maple Firemist	3532(Cad.)
91	Lt.Amethyst Firemist	3997(B,O,Cad.)
92	Med. Emerald Firemist	3904(B,O)
96	Dark Driftwood Firemist	3906(B,O,Cad.)
98	Chamois Firemist	3908(B,O,Cad.)
99	Dark Amethyst Firemist	3998(B,O,Cad.)

B-Buick, C-Chevrolet, O-Oldsmobile, P-Pontiac,
Cad.-Cadillac, Cam-Camaro, Cor-Corvette, F'Bd-
Firebird

1988 GENERAL MOTORS

11	White	2058(B,C,O,P,Cad.)
12	Silver Poly	3746(B,C,O,P,Cad.)
12C	Silver Poly	4081(B,C,O,P)
13	Silver Poly	3661(C,Cor.)
13C	Silver Poly	4128(Cam,F'bd,T)
13S	Silver Poly	3822(B,C,O,P,Cad.)
14	Black Poly	3961(B,C,O,P,Cad.)
15	Med. Gray Poly	3749(B,C,O,P,Cad.)
15C	Med. Gray Poly	3921(B,C,O,P)
16	Silver Poly	3880(P)
17	Lt. Pearl Gray	3963(Cad.)
17	Lt. Pearl Gray	4083(B,O,Cad.)
19	Black	9300(B,C,O,P,Cad.)
20	Nassau Blue Poly	3881(C,Cor.)
22	Lt.Sapphire Blue Poly	3964(B,C,O,P,Cad.)
23	Med. Maui Blue Poly	3882(Cam,F'Bd)

23	Maui Blue Poly	4129(Cam,F'Bd,T)
24C	Bright Blue Poly	4106(B,C,O,P,Cad.)
25	Lt. Blue Poly	3754(B,C,O,P,Cad.)
26	Lt. Saphire Blue Poly	3966(B,C,O,P,Cad.)
26C	Lt. Saphire Blue Poly	4085(B,C,O,P,Cad.)
27	Med.Sapphire Blue Poly	3967(B,C,O,P,Cad.)
28	Black Sapphire Poly	3885(B,C,O,P,Cad.,Cor.)
29	Dk. Saphire Blue Poly	3968(B,C,O,P,Cad.)
29C	Dk. Saphire Blue Poly	4086(B,C,O,P,Cad.)
31	Dk. Blue Poly	3760(B,C,O,P)
32	Black Sapphire Firemist	4087(Cad.)
33	Lt. Driftwood Poly	3970(Cor.)
40	White	3680(B,C,O,P,Cad.,Cor.)
41	Black	9700(B,C,O,P,Cad.,Cor.)
42	Lt. Emerald Poly	3973(B,C,P,)
50	Flax	3529(Cad.)
51	Yellow Gold	3889(Cam,F'Bd)
52	Copper Beige	4088(B,C,O,P,Cad)
52	Copper Beige	3976(B,C,O,P,Cad)
53	Talbot Yellow	4089(P)
57	Lt. Beechwood Poly	4090(B,C,O,P,Cad.)
57C	Lt. Beechwood Poly	4091(B,C,O,P,Cad.)
58	Lt. Chestnut Poly	3772(B,C,O,P)
59	Dk. Beechwood Poly	4092(B,C,O,P)
59C	Dk. Beechwood Poly	4093(B,C,O,P)
62	Lt. Beechwood Poly	4094(B,C,O,P,Cad.)
63	Med. Orange Poly	4095(Cam,F'bd)
63	Orange Poly	4130(Cam,F'bd)
66	Dk. Orange Poly	4008(Cor.T)
67	Dk. Antelope Poly	4096(B,O,Cad.)
69	Med. Beechwood Poly	4109(B,O,P,Cad.)
70	Carmine	3781(B,C,O,P,Cad)
72	Med. Garnet Red Poly	3986(B,O,Cad.)
73	Brilliant Red	4098(B,O,P)
74	Flame Red Poly	3823(Cam,Cor,F'Bd)
74	Flame Red Poly	4131(Cam,Cor,T)
75	Med.Garnet Red Poly	3988(B,C,O,P)
75C	Med.Garnet Red Poly	4099(B,C,O,P)
76	Dk.Garnet Red Poly	3989(B,O,P,Cad.)
77	Med. Red Poly	3990(P)
78	Med. Rosewood Poly	3991(B,C,O,P)
78C	Med. Rosewood Poly	4100(B,C,O,P)
79	Dk. Red Poly	3786(B,C,O,P,Cad.)
80	Lt. Rosewood Poly	3993(B,O,Cad.)
81	Bright Red	3794(Cam,Cor.F'bd,)
82	Med.Rosewood Poly	3994(B,O,Cad.)
83	Med. Sapphire Blue	3995(B,O,Cad.)

	F.M.	
86	Antelope Beige	4101(B,O,Cad.)
87	Gunmetal Poly	4009(B,C,O,P,Cad.)
87	Gunmetal Poly	4132(Cam,F'Bd)
88	Dark Gray Poly	3788(Cad.)
89	Autumn Maple Poly	3532(Cad.)
89C	Autumn Maple Poly	4102(B,C,O,P)
90	Med. Smoke Gray Poly	4103(Cor.)
95	Lt. Antelope Firemist	4104(B,O,Cad.)
96	Dark Smoke Gray Poly	4105(Cor.)

B-Buick, C-Chevrolet, O-Oldsmobile, P-Pontiac,
Cad.-Cadillac, Cam-Camaro, Cor-Corvette,
F'Bd-Firebird, T-Trans AM

1989 GENERAL MOTORS

10	Arctic White	4185(Cor.)
11	White	2058(B,C,O,P,Cad.)
12	Silver Poly	3746(B,C,O,P,Cad.)
12C	Silver Poly	4081(B,C,O,P)
13S	Silver Poly	3822(B,C,O,P)
14	Black Poly	3961(B,C,O,P)
15	Med. Gray Poly	3749(B,C,O,P,Cad.)
I5C	Med. Gray Poly	4082(B,C,O,P)
17	Lt. Pearl Gray	3963(Cad.)
17	Lt. Pearl Gray	4083(B,O,)
19	Black	9000/9300(B,C,O,P, Cad.)
20	Nassau Blue Poly	3881(Cor.)
22	Lt.Sapphire Blue Poly	3964(B,C,O,P,Cad.)
23	Med. Maui Blue Poly	4129(C,P)
24	Bright Blue Poly	3883(B,C,O,P)
26	Lt.Sapphire Blue Poly	3966(B,C,O,P,Cad)
26C	Lt. Sapphire Blue Poly	4085(B,C,O,P)
27	Med. Sapphire Blue Poly	3967(B,C,O,P)
28	Black Saphire Poly	3885(B,C,O,P,Cor.)
29	Dk. Sapphire Blue Poly	3968(B,C,O,P,Cad.)
29C	Dk. Sapphire Blue Poly	4086(B,C,O,P)
31	Arctic Pearl	4176(Cor.)
32	Black Saphire Poly	4087(Cad.)
33C	Light Driftwood Poly	3970(C)
35	Yellow	3886(Cor.)
40	White	3680(B,C,O,P)
41	Black	9700(B,C,O,P)
50	Flax	3668(B)
52	Copper Beige	3976(B,O)

52	Copper Beige	4088(B,C,O,P,Cad.)
54	Med.Flax	4133(Cad.)
57	Lt. Beechwood Poly	4090(B,C,O,P.Cad.)
57C	Lt. Beechwood Poly	4091(B,C,O,P)
59	Dk. Beechwood Poly	4092(B,C,O,P)
59C	Dk. Beechwood Poly	4093(B,C,O,P)
62	Lt. Beechwood Poly	4094(B,C,O,P,Cad.)
67	Dk. Antelope Poly	4096(B,O)
68	Brilliant Red Poly	4177(C,Cor.)
69	Med. Beechwood Poly	4109(B,O,P,Cad.)
70	Carmine	3781(B,C,O,P)
70C	Carmine	4097(B,C,O,P)
72	Med. Garnet Red Poly	3986(B,C,O,P)
73	Brilliant Red	4098(B,O,P)
74	Flame Red Poly	4131(C,P)
75	Med. Garnet Red Poly	3988(B,C,O,P)
75C	Med. Garnet Red Poly	4099(B,C,O,P)
76	Dk. Garnet RedPoly	3989(B,O,P)
78	Med. Rosewood Poly	3991(B,C,O,P)
78C	Med. Rosewood Poly	4100(B,C,O,P)
79	Dk. Red Poly	3786(B,C,O,P,Cad.)
81	Bright Red	3794(B,C,O,P,Cor.)
82	Med. Rosewood Poly	3994(B,O)
83	Med. Sapphire Blue Firemist	3995(B,O,P)
86	Antelope Beige	4101(B,O,Cad.)
87	Gunmetal Poly	4009(B,C,O,P)
87	Gunmetal Poly	4132(C,P)
89	Autumn Maple Poly	3532(B,C,O,P,Cad.)
89C	Autumn Maple Poly	4102(B,C,O,P)
90	Med. Smoke Gray Poly	4103(Cor.)
93	White Diamond-Base	3680(Cad.)
93	White Diamond-Top	4175(Cad.)
95	Lt. Antelope F.M.	4104(B,O)
96	Dk. Smoke Gray Poly	4105(Cor.)
98	Bright Blue Poly	4136(C,P)

B-Buick, C-Chevrolet, O-Oldsmobile, P-Pontiac,
Cad.-Cadillac, Cor-Corvette,

1990 GENERAL MOTORS

08	Mary Kay Pink	70748(Special Order)
08	Mary Kay Pink	3365(Special Order)
10	Artic White	4185(Cor.)
12	Silver Poly	3746(B,C,O,P,Cad.)
12	Silver Poly	4081(B,C,O,P)
13	Silver Poly	4128(Special Order)

13S	Silver Poly	3822(B,C,O,P,Cad.)
14	Black Poly	3961(B,C,O,P,Cad.)
15	Med. Gray Poly	3749(B,C,O,P,Cad.)
15	Med. Gray Poly	4082(B,C,O,P,Cad.)
17	Lt. Pearl Gray	3963(Cad.)
18	Slate Gray	4222(B,O,Cad.)
20	Nassau Blue Poly	3881(Cor.)
21	Neon Blue Poly	4268(B,C,O,P)
22	Lt. Saphire Blue Poly	3964(B,C,O,P,Cad.)
23	Med. Maui Blue Poly	4084(B,O)
23	Med. Maui Blue Poly	4129(C,P)
24	Bright Blue Poly	3883(B,C,O,P)
24	Bright Blue Poly	4106(B,C,O,P)
25	Steel Gray Poly	4223(Cor.)
26	Lt. Sapphire Blue Poly	3966(B,C,O,P,Cad.)
26	Lt. Sapphire Blue Poly	4085(B,C,O,P)
27	Med. Sapphire Blue Poly	3967(B,C,O,P)
28	Black Sapphire Poly	3885(B,C,O,P,Cad.)
29	Dk. Sapphire Blue Poly	3968(B,C,O,P,Cad.)
29	Dk. Sapphire Blue Poly	4086(B,C,O,P)
32	Black Sapphire Firemist	4087(Cad.)
33	Lt. Driftwood Poly	3970(B)
40	White	3680(B,C,O,P,Cad.)
40	White	2058(B,C,O,P,Cad.)
41	Black	9000/9300(B,C,O,P,41 Cad.)
	Black	9700(B,C,O,P,Cad.,Cor.)
42	Warm Silver	34607(B,O,Cad.)
42	Turquoise Poly	4265(C,O,P)
44	Malachite Poly	4224(C)
50	Flax	3668(Cad.)
52	Copper Beige	4088(B,C,O.P,Cad.)
53	Competition Yellow	4266(Cor.)
54	Med. Flax	4133(Cad.)
57	Camel Beige	4225(B,C,O,P)
57	Lt.Beechwood Poly	4090(B,C,O.P,Cad)
58	Charcoal	34249(C)
59	Dk. Beechwood Poly	4092(B,C,O,P)
59	Dk. Beechwood Poly	4093(B,C,O,P)
62	Lt. Beechwood Poly	4094(B,O,P)
64	Lt. Auburn Poly	4226(B,O,Cad)
65	Dk. Sable Poly	4252(Cad.)
65	Lt. Camel Titanium	4264(B,O,C)
66	Dk. Auburn Poly	4227(B,O,Cad)
68	Brilliant Red Poly	4177(Cor.)

69	Med. Beechwood Poly	4109(B,O)
70	Torch Red	4228(B,C,O,P)
70	Torch Red	4229(B,C,O,P)
71	Torch Red	4230(C)
72	Med.Garnet Red Poly	3986(B,C,O,P,Cad)
74	Flame Red Poly	3823(B,O,P)
75	Brilliant Red Poly	4231(C,P)
76	Dk. Garnet Red Poly	3989(B,O,P,Cad.)
78	Dk. Maple Poly	4232(B,C,O)
79	Dk. Maple Poly	4233(B,C,O,P,Cad.)
80	Med. Quasar Blue Poly	4267(Cor.)
81	Bright Red	3794(B,C,O,P,Cor.)
83	Med. Saphire Blue F.M.	3995(B,O,Cad.)
85	Med. Slate Poly	4234(B,C,O,P,Cad.)
87	Gunmetal Poly	4009(B,C,O,P)
87	Gunmetal Poly	4132(C,P)
89	Autumn Maple Poly	3532(B,C,O,P,Cad.)
89	Autumn Maple Poly	4102(B,C,O,P)
91	Polo Green Poly	4235(Cor.,Cad.)
93	White Diamond Base	3680(B,Cad.)
93	White Diamond Pearl	4175(B,Cad.)
94	Lt. Antelope Poly	4236(Cad.)
95	Lt. Antelope Firemist	4104(B,O.Cad.)
96	Dk. Smoke Gray Poly	4105(Cor.)
98	Bright Blue Poly	4136(C,P)
99	Dk. Slate Gray Poly	4237(Cad.)

B-Buick, C-Chevrolet, O-Oldsmobile, P-Pontiac, Cad.-Cadillac, Cor-Corvette,

1991 GENERAL MOTORS

08	Mary Kay Pink	70748(Special Order)
08	Mary Kay Pink	3365(Special Order)
10	Artic White	4185(P,C)
12	Silver Poly	3746(B,C,P)
12	Silver Poly	4081(B,C,O,P)
13	Bright Silver Poly	3822(B,C,O,P,Cad.)
15	Med. Gray Poly	3749(B,C,P)
15	Med. Gray Poly	4082(B,C,O,P)
16	Brite White	4298(P)
17	Pearl Gray	4083(Cad.)
18	Slate Gray	4222(B,O,Cad.)
21	Neon Blue Poly	4268(P)
22	Lt. Saphire Blue Poly	3964(B,C,O,P,Cad.)
22	Lt. Saphire Blue Poly	4085(B,C,P)
23	Med. Maui Blue Poly	4084(B,O)
23	Med. Maui Blue Poly	4129(C,P)
24	Bright Blue Poly	3883(B,C,P)
24	Bright Blue Poly	4106(B,C,P)

25	Steel Gray Poly	4223(Cor.)
27	Med. Sapphire Blue Poly	3967(B,C,O,P)
28	Black Sapphire Poly	3885(B,C,O,P,Cad.)
29	Dk. Sapphire Blue Poly	3968(B,C,P)
29	Dk. Sapphire Blue Poly	4086(B,C,O,P)
33	Lt. Driftwood Poly	3970(B)
35	Yellow	3886(B,C,P)
37	Dk. Bright Teal Poly	4300(C)
40	White	4317(B,C,P)
40	White	4318(B,C,O,P,Cad.)
41	Black	9000/9300(B,C,P)
41	Black	9700(B,C,O,P,Cad.,Cor.)
42	Turquoise Poly	4265(C,P,Cor.)
44	Malachite Poly	4224(C)
45	Med. Green Poly	4301(P)
50	Flax	3668(Cad.)
51	Dk. Antelope Poly	4096(Cad.)
53	Competition Yellow	4266(Cor.)
54	Med.Flax	4133(Cad.)
56	Alabaster	4302(B)
57	Camel Beige	4225(B,C,O,P)
62	Lt. Beechwood Poly	4094(B,O,P)
64	Lt. Auburn Poly	4226(B,O,Cad.)
66	Dk. Auburn Poly	4227(B,O,Cad.)
67	Lt. Camel Poly	4303(B,C,P)
69	Med. Beechwood Poly	4109(B,O)
70	Torch Red	4230(C)
72	Med.Garnet Red Poly	3986(B,C,O,P,Cad)
74	Flame Red Poly	3823(B,C,O,P,Cad.)
75	Brilliant Red Poly	4231(C,P)
75	Brilliant Red Poly	4177(Cor.)
76	Dk. Garnet Red Poly	3989(B,O,P,Cad.)
78	Dk. Maple Poly	4232(B,C,O,P.Cad.)
80	Med. Quasar Blue Poly	4267(Cor.)
81	Bright Red	3794(B,C,O,P,Cad,Cor.)
83	Med. Saphire Blue F.M.	3995(B,O,Cad.)
85	Med. Slate Poly	4234(B,C,O,P,Cad.)
87	Gunmetal Poly	4009(B,C,O,P)
87	Gunmetal Poly	4132(C,P)
89	Autumn Maple Poly	3532(B,C,P)
89	Autumn Maple Poly	4102(B,C,P)
91	Dk. Polo Green Ivy	4235(Cor.,Cad.)
93	White Diamond Base	3680(B,Cad.)
93	White Diamond Pearl	4175(B,Cad.)
95	Lt. Antelope Firemist	4104(B,O.Cad.)

96	Dk. Smoke Gray Poly	4105(Cor.)
98	Ultra Blue Poly	4136(C,P)
99	Dk. Slate Gray Poly	4237(Cad.)

B-Buick, C-Chevrolet, O-Oldsmobile, P-Pontiac, Cad.-Cadillac, Cor-Corvette,

1992 GENERAL MOTORS

08	Mary Kay Pink	73365
09	Mary Kay Pink Pearlcoat Base	73255/Top 4170
09	Opaque White	91013(A)
10	Arctic White Solid	4185(C,P,COR.)
13	Bright Silver Poly	3822(B,C,O,P,S,CAD.)
14	Light Gray Poly	4452(C,P)
15	Medium Gray Poly	4082(B,C,O,P,S)
16	Brite White Solid	4298(B,C,O,P)
18	Dark Green Gray Poly	4453(All)
20	Saturn Blue Poly	4332(S)
21	Neon Blue Poly	4268(B,C,O,P)
22	Light Sapphire Blue Poly	3984(B,C,O,P)
23	Medium Maui Blue Poly	4084(B,C,P)
25	Aquamarine Poly	4532(S)
27	Med. Sapphire Blue Poly	3967(P)
28	Black Sapphire Poly	3885(B,C,O,P,CAD)
28	Black Sapphire Poly	4334(S)
29	Dark Sapphire Blue Poly	4086(P)
30	Specular Silver Poly	4010(A)
33	Light Driftwood Poly	4454(B,C,O,P,CAD)
35	Yellow Solid	3386(COR)
37	Dark Bright Teal Poly	4300(C,P)
38	Champagne Poly	4263(A)
39	Dark Blue Green Poly	4335(S)
40	White Solid	4318(B,C,O,P,CAD)
40	White	4336(S)
41	Black	9700(All)
41	Black	9729(All)
41	Black	4337(S)
42	Turquoise Poly	4265(C)
43	Bright Aqua Poly	4456(B,C,O,P,CAD)
44	Medium Malchite Poly	4224(C,O,P)
45	Medium Green Poly	4301(C,P,COR)
47	Euro Red	4125(A)
48	Dark Yellow Green Poly	4457(B,O,P)
49	Light Blue Poly	4191(All)
51	Dark Antelope Poly	4096(S,CAD)

52	Jamaica Yellow Solid	4458(P)
55	Light Canyon Yellow	4459((B,O,P,CAD)
55	Euro White Base	4013/Top 4014(A)
56	Alabaster Solid	4302(B)
57	Light Beige Poly	4460(B,C,O,P,CAD)
58	Canyon Yellow Tri-Coat Base	4461/Top 4528(CAD)
63	Medium Taupe Poly	4462(CAD)
67	Light Camel Poly	4303(B,C,O,P,)
67	Light Camel Poly	4304(B,C,P)
69	Rootbeer Poly	4463(B,O,CAD)
70	Torch Red	4230(C)
72	Medium Garnet Red Poly	3986(B,C,O,P,CAD)
73	Blackrose Poly	4464(COR)
74	Flame Red Poly	3823(B,C,O,P,CAD)
75	Brilliant Red Poly	4231(C,P)
75	Brilliant Red Poly	4177(COR)
75	Brilliant Red Poly	4340(S)
76	Dark Garnet Red Poly	3989(B,C,O,P,CAD)
77	Dark Cherry Poly	4465(CAD)
78	Dark Maple Red Poly	4232(B,C,O,P,CAD)
79	Burgundy Poly	4127(All)
80	Med. Quasar Blue Poly	4267(B,C,O,P,COR)
81	Bright Red	3784(B,C,O,P,COR)
81	Bright Red	4341(S)
83	Med. Sapphire Blue Firemist	3995(O,CAD)
84	Hawaiian Orchid Poly	4466(C)
85	Medium Slate Poly	4234(B,C,O,CAD)
87	Gunmetal Poly	4009(B,C,O,CAD)
88	Dark Plum Poly	4467(CAD)
89	Autumn Maple Poly	4102(B,C,P)
91	Dark Polo Green Poly	4235(CAD)
93	White Diamond Tr-Coat Base	3680/Top 4175(CAD)
95	Light Antelope Firemist	4104(CAD)

A-Allante, B-Buick, C-Chevrolet, O-Oldsmobile,
P-Pontiac, Cad.-Cadillac, S-Saturn, Cor-Corvette

1993 GENERAL MOTORS

04	Pearl Flax	4550/4551(A)
09	Mary Kay Pink Pearlcoat Base	73255/Top 4170(A)
09	Opaque White	91013(A)
10	Arctic White Solid	4185(C,COR.)
13	Bright Silver Poly	3822(B,C,O,P,S,CAD.)
14	Light Gray Poly	4452(C,O,P)
15	Medium Gray Poly	4082(B,O,S)

16	Bright White	4298(B,C,O,P)
18	Dark Green Gray Poly	4701(C)
20	Saturn Blue Poly	4332(S)
21	Neon Blue Poly	4268(P)
22	Light Sapphire Blue Poly	3964(B,C,O,P,CAD)
23	Medium Maui Blue Poly	4084(B,C)
25	Aquamarine Poly	4532(S)
27	Med. Sapphire Blue Poly	3967(P)
28	Black Sapphire Poly	3885(B,O,P,CAD)
28	Black Sapphire Poly	4334(S)
29	Dark Sapphire Blue	4086(P)
30	Med. Adriatic Blue Poly	4702(B,C)
33	Light Driftwood Poly	4454(B,C,O,P,CAD)
33	Beige Poly	4723(S)
34	Light Teal Poly	4703(B,C,P)
38	Dark Red Poly	4724(S)
38	Champagne Poly	4263(A)
39	Dark Blue Green Poly	4335(S)
40	White	4318(B,C,CAD)
40	White	4336(S)
41	Black	9700(C,P,CAD)
41	Black	9729(All)
41	Black	4337(S)
43	Bright Aqua Poly	4456(C,O,P)
44	Medium Malchite Poly	4224(B,C,O,P)
45	Medium Green Poly	4301(C,P,COR)
47	Euro Red	4126(A)
48	Dark Yellow Green Poly	4457(C,P)
52	Jamaica Yellow Solid	4458(P)
53	Competition Yellow	4266(COR)
55	Euro White Base	4013/Top 4014(All)
56	Alabaster Solid	4302(B,CAD)
57	Light Beige Poly	4460(B,O,P,CAD)
58	Canyon Yellow Pearlcoat	4705/4706(CAD)
59	Bronze Poly	4707(CAD)
63	Medium Taupe Poly	4462(CAD)
68	Dark Ruby Poly	4708(COR)
69	Rootbeer Poly	4463(B)
70	Torch Red Solid	4230(B,C,O,P)
71	Medium Patriot Red Poly	4709(C,P)
72	Medium Garnet Red Poly	3986(B,C,O,P,CAD)
73	Blackrose Poly	4464(COR)

102

75	Brilliant Red Poly	4177(COR)
75	Brilliant Red Poly	4340(S)
76	Dark Garnet Red Poly	3989(B,C)
77	Dark Cherry Poly	4465(B,O,CAD)
80	Med. Quasar Blue Poly	4710(C,P)
80	Med. Quasar Blue Poly	4267(B,C,O,P,COR)
81	Bright Red Solid	3794(B,C,O,P,CAD)
81	Bright Red	4341(S)
83	Med. Sapphire Blue Firemist	3995(B,O,CAD)
84	Pearl Red Tri-Coat	4553/4554(All)
87	Gunmetal Poly	4009(B,O,CAD)
88	Dark Plum Poly	4467(CAD)
89	Autumn Maple Poly	4102(B,C,P)
91	Dark Polo Green Poly	4235(A,CAD)
93	White Diamond Tri-Coat	4712/4713(CAD)

A-Allante, B-Buick, C-Chevrolet, O-Oldsmobile, P-Pontiac, Cad.-Cadillac, S-Saturn, Cor-Corvette

1994 GENERAL MOTORS

09	Mary Kay Pink	73255(B,C,O,P,CAD)
10	Arctic White Solid	4185(C)
13	Bright Silver Poly	3822(B,O,CAD.)
13	Bright Silver Poly	4330(S)
14	Light Gray Poly	4452(B,O,P,CAD)
15	Medium Gray Poly	4331(S)
16	Bright White Solid	4298(B,C,O,P,CAD)
18	Dark Green Gray Poly	4453(B,C,O,P,CAD)
18	Dark Green Gray Poly	4701(C)
20	Saturn Blue Poly	4332(S)
21	Med. Montana Blue Poly	4794(B,O,CAD)
22	Dark Montana Blue Poly	4795(B,O,CAD)
23	Light Cloisonne Poly	4796(B,C,O,P)
25	Aquamarine Poly	4532(S)
26	Med. Cloisonne Poly	4797(B,C,O,P)
28	Dark CLoisonne Poly	4798(COR)
30	Med. Adriatic Blue Poly	4702(B,C,O,P,CAD)
33	Light Driftwood Poly	4454(B,C,O,P,CAD)
33	Beige Poly	4723(S)
34	Light Teal Poly	4703(C,P)
34	Light Teal Poly	4800(B,O)
35	Dark Teal Poly	4801(B,C,O,P)
36	Medium Blue Green	4858(S)
36	Light Adriatic Poly	4812(B,C,O,P,CAD)
36	Light Adriatic Poly	4802(B,O,P,CAD)
37	Dark Red Poly	4724(S)

37	Med. Teal Blue Poly	4803(C,P)
39	Dark Blue Green Poly	4335(S)
39	Dark Adriatic Poly	4805(B,C,O,P,CAD)
40	White	4318(B,C,CAD)
41	Black	9835(B,C,O,P,CAD, COR)
41	Black	4337(S)
43	Bright Aqua Poly	4456(B,C,O,P,COR)
43	Bright Aqua Poly	4859(B,C,O,P)
44	Medium Malachite Poly	4224(B,C,O,P)
45	Medium Green Poly	4301(COR)
48	Dark Yellow Green Poly	4457(B,C,O,P)
53	Competition Yellow Solid]	4266(COR)
54	Sunfire Yellow Solid	4704(C,P)
57	Light Beige Poly	4460(B,O,P,CAD)
60	Mocha Poly	4807(B,O,CAD)
63	Medium Taupe Poly	4462(CAD)
66	Melon Poly	4808(COR)
70	Torch Red Solid	4230(B,C,O,P,COR)
71	Medium Patriot Red Poly	4709(C,P)
72	Medium Garnet Red Poly	3986(B,C,O,P,CAD)
73	Black Rose Poly	4464(C,COR)
75	Brilliant Red Poly	4177(COR)
75	Brilliant Red Poly	4340(S)
76	Dark Garnet Red Poly	3989(B,C,O,P)
77	Dark Cherry Poly	4465(B,C,O,P,CAD)
80	Med. Quasar Blue Poly	4710(C,P)
81	Bright Red Solid	3794(B,C,O,P)
81	Bright Red	4341(S)
83	Light Antelope Poly	4809(B,O)
84	Hawaiian Orchid Poly	4466(C,P)
87	Gunmetal Poly	4009(B,C,O,CAD)
89	Medium Mushroom Poly	4810(B,O)
91	Dark Polo Green Poly	4235(B,CAD)
91	Dark Polo Green Poly	4711(B,C,O,P,CAD)
92	Dark Calypso Green Poly	4811(B,O,CAD)
93	White Diamond	3680/4175(B,O,CAD)

B-Buick, C-Chevrolet, O-Oldsmobile, P-Pontiac, Cad.-Cadillac, S-Saturn, Cor-Corvette

1995 GENERAL MOTORS

| 04 | Medium Marblehead Poly | 4991(B,C,O,P,CAD) |
| 05 | Cyclamen Poly | B,C,O,P,COR) |

89	Mary Kay Pink	73365(B,C,O,P,CAD)
10	Arctic White Solid	4185(C,P,COR))
13	Bright Silver Poly	3822(B,C,O,P)
13	Bright Silver Poly	4330(C,P)
14	Light Gray Poly	4452(B,C,O,P,CAD)
16	Bright White Solid	4298(B,C,O,P,CAD)
18	Dark Green Gray Poly	4453(B,C,O,P,CAD)
18	Dark Green Gray Poly	4701(C)
20	Light Montana Blue Poly	4792(CAD)
21	Med. Montana Blue Poly	4794(CAD)
22	Dark Montana Blue Poly	4795(CAD)
26	Med. Cloisonne Poly	4797(B,C,O,P)
28	Dark CLoisonne Poly	4798(COR)
30	Med. Adriatic Blue Poly	4702(B,C,O,P,CAD)
31	Medium Teal Poly	4799(B,C,O,P)
33	Light Driftwood Poly	4454(B,C,O,P,CAD)
34	Light Teal Poly	4800(B,O)
35	Dark Teal Poly	4801(B,C,O,P)
36	Light Adriatic Poly	4812(B,C,O,P,CAD)
36	Light Adriatic Poly	4802(B,O,P,CAD)
37	Med. Teal Blue Poly	4803(C,P)
39	Dark Adriatic Poly	4805(B,O,CAD)
40	White	4318(CAD)
41	Black	9835(B,C,O,P,CAD, COR)
43	Bright Aqua Poly	4456(B,C,O,P,COR)
43	Bright Aqua Poly	4859(B,C,O,P)
44	Medium Malachite Poly	4224(B,C,O,P)
45	Medium Green Poly	4301(COR)
48	Dark Yellow Green Poly	4457(B,C,O,P)
46	Shale Poly	4995(CAD)
53	Competition Yellow Solid	4266(COR)

54	Sunfire Yellow Solid	4704(C,P)
55	Light Autumnwood Poly	4996(C,P)
57	Light Beige Poly	4460(B,O,P)
60	Mocha Poly	4807(CAD)
70	Torch Red Solid	4230(B,C,O,P,COR)
71	Medium Patriot Red Poly	4709(C,P)
72	Medium Garnet Red Poly	3986(B,C,O,P,CAD)
73	Black Rose Poly	4464(C,COR)
75	Brilliant Red Poly	4177(COR)
77	Dark Cherry Poly	4465(B,C,O,P,CAD)
80	Med. Quasar Blue Poly	4710(C,P)
81	Bright Red Solid	3794(B,C,O,P)
82	Dark Red	4998(B,C,O,P)
83	Light Antelope Poly	4809(B,O)
84	Hawaiian Orchid Poly	4466(C,P)
86	Red Tint Coat Poly Tri-Coat	4999/5000(B,O)
89	Medium Mushroom Poly	4810(B,O)
91	Dark Polo Green Poly	4235(B,CAD)
91	Purple Pearl	4711(B,C,O,P,CAD)
92	Dark Calypso Green Poly	4811(CAD)
93	White Diamond	3680/4175(CAD)
95	Majestic Amethyst Poly	5001(CAD)
96	Cayenne Red Poly	5002(B,C,O,P)
98	Raspberry Poly Orange	5003(C,P) (COR)

B-Buick, C-Chevrolet, O-Oldsmobile, P-Pontiac, Cad.-Cadillac, S-Saturn, Cor-Corvette

Chapter 10

LINCOLN 1941-1978

PAINT CODE	COLOR	PPG CODE
1941 LINCOLN-ZEPHYR		
M-13005	Seamist Poly (Upper)	30105
	Darian Blue Poly(Lower)	10114
M-1306	Staffordshire Green (Upper)	40120
	Beetle Green (Lower)	40076
M-1307	Curacao Brown (Upper)	20076
	Volanta Coach Maroon (Lower)	50027
	Cape CodeGray (Upper)	30104
	Zephyr Blue (Lower)	10073
	Black	9000
	Beetle Green	40076
	Volanta Coach Maroon	50027
	Darian Blue Poly	10114
	Rockingham Tan	20077
	Paradise Green	40022
	Plympton Gray Poly	30107
	Capri Blue Poly	10060

PAINT CODE	COLOR	PPG CODE
1942 LINCOLN		
	Black	9000
	Darian Blue Poly	10100
	Chetwyn Beige	20056
	Andover Green	40058
	Bristol Blue	10099
	Victoria Coach Maroon	50022
	Suwanee Green	40059
	Sheldon Gray Poly	30081

PAINT CODE	COLOR	PPG CODE
1946 LINCOLN		
	Black	9000
	Lincoln Maroon	50042
	Sheldon Green Poly	30187
	Skyling Blue Poly	10174
	Surf Green Poly	40183
	Willow Green	40184
	Marine Blue	10173
	Wing Gray	30192

PAINT CODE	COLOR	PPG CODE
1947 LINCOLN		
	Black	9000
	Regal Blue	10378
	Grotto Blue	10379
	Dune Beige	20041
	Canyon Tan	20385

	Sea Gull Gray	30416
	Steel Gray Poly	30417
	Valley Green Poly	40466
	Opal Blue Green	40467
	Moss Green	40469
	Lincoln Maroon	50042
	Pace Car Yellow	80288

PAINT CODE	COLOR	PPG CODE
1949 LINCOLN-MERCURY		
8L,M	Biscay Blue Poly	10474
8H	Biscay Blue Poly	10491
8H	Alberta Blue	10479
8L,M	Alberta Blue	10480
8H	Teal Blue	10481
M	Alaska Blue Gray Poly	10445
8H	Haiti Beige	20463
8L,M	Haiti Beige	20446
8L,M	Lima Tan Poly	20470
8H	Lima Tan Poly	20471
8H	Temple Gray	30465
8H	Dakota Gray	30522
8L,M	Dakota Gray	30478
8H	Mogul Green Gray	30483
8H	Blue Steel Gray Poly	30537
8L	Hampton Gray Poly	30472
M	Cairo Gray	30493
8H	Adalia Green	40587
8L,M	Banff Green Poly	40531
8H	Banff Green Poly	40595
8L,M	Berwick Green Poly	40546
8L	Lido Green Poly	40398
8H	Lido Green Poly	40570
8H	Calcutta Green	40532
8L,M	Royal Bronze Maroon Poly	50126
8H	Royal Bronze Maroon Poly	50131
M	Midland Maroon Poly	50121
8L,M	Tampico Red Poly	70245
8H	Pirate Red	70224
8H	Lincoln Ivory	80360
M	Bermuda Cream	80177
8H	Calabash Yellow	80345
8L	Calabash Yellow	80363
8H,8L,M	Black	9000

MODEL CODES 8H = LINCOLN 8H 8L =
LINCOLN 8L M = MERCURY

TWO-TONES
8L,M U 20470   8H U 40587   8L,M U 40531

	L 20446		L 30483		L 40546
8H	U 20471	8H	U 30465		
	L 20463		L 30522		

1950 LINCOLN

110	Haiti Beige	20544
111	Admiral Blue	10588
112	Cosmopolitian Maroon Poly	50162
113	Newport Gray	30657
114	Nassau Beige Poly	20590
114X	Nassau Beige Poly	20688
115	Mallard Green Poly	40742
115X	Mallard Green Poly	40887
116	Arrowhead Gray Poly	30656
116X	Arrowhead Gray Poly	30722
117	Danube Blue Poly	10589
117X	Danube Blue Poly	10661
118	Palomar Green Poly	40741
118X	Palomar Green Poly	40886
119	Carlsbad Tan	20589
120	Chantilly Green	40735

TWO-TONES

122	U 30722	125	U 30657	128	U 20668
	L 30609		L 30722		L 20589
123	U 30657	126	U 30722	129	U 40748
	L 10661		L 30657		L 40886
124	U 10661	127	U 20589	130	U 30609
	L 30657		L 20668		L 30722

1951 LINCOLN

01	Raven Black	9000
02	Admiral Blue Poly	10588
02A	Admiral Blue (Solid)	10148
03	Banning Blue Poly	10582
04	Cosmopolitan Maroon Poly	50162
04A	Cosmopolitan Maroon (Solid)	50030
05	Luxor Maroon Poly	50218
06	Kent Gray (Blue)	10771
08	Copper Tone Poly	20764
08A	Copper Tone (Solid)	20086
09	Tomah Ivory	80526
10	Brewster Green Poly	40976
10A	Brewster Green (Solid)	40286
11	Everglade Green	40894
12	Saxon Gray Poly	30863
12A	Saxon Gray (Solid)	30150
13	Sheffield Green	40944
14	Avon Blue	10769
16	Radiant Green Poly	40975
18	Bristol Buff	20765
20	Chantilly Green	40735
21	Manitou Red Poly	50217
21A	Manitou Red (Solid)	50038

TWO-TONES

30	U 9000	44A	U 30150	53	U 20765
	L 80526		L 10771		L 20764
31	U 80526	45	U 10588	53A	U 40976
	L 40944		L 10771		L 20086
32	U 40944	46	U 10771	206	U 40976
	L 80526		L 10588		L
33	U 30856	46A	U 10771	206AU	
	L 40894		L 10148		L 40977
34	U 40894	47	U 40975	206B	U
	L 30856		L 40976		L 40286
35	U 40944	47A	U 40975	207	U
	L 40894		L 40286		L 10772
36	U 40894	48	U 40976	208	U
	L 40944		L 40975		L 9000
41	U 10771	48A	U 40286	210	U
	L 10769		L 40975		L 80530
42	U 10769	49	U 40975	211AU	
	L 10771		L 20765		L 50038
43	U 10771	50	U 20765	212AU	
	L 30863		L 40975		L 10148
43A	U 10771	51	U 10582		
	L 30150		L 10771		
44	U 30863	52	U 10771		
	L 10771		L 10582		

1952 LINCOLN

01A	Raven Black	9000
02A	Admiral Blue	10820
04A	Fanfare Maroon	50287
05A	Newport Gray	20086
08	Coppertone	20886
08A	Coppertone Poly	20797
09A	Pebble Tan	20489
10A	Academy Blue	10011
11A	Hillcrest Green	41131
13A	Lakewood Green	41107
15	Saxon Gray	30150
15A	Saxon Gray Poly	30912
18A	Cinabar Red	70382
19A	Vassar Yellow	80534

TWO-TONES

22	U 20489	33A	U 50287	44A	U 30909
	L 20086		L 20489		L 10011
22A	U 20489	34A	U 9000	45A	U 10011
	L 20797		L 80534		L 30909

23	U 20086 / L 20489	36	U 30909 / L 30150	46A	U 9000 / L 20489
23A	U 20797 / L 20489	36A	U 30909 / L 30912	49A	U 30909 / L 10820
26A	U 41131 / L 30909	37	U 30150 / L 30909	50A	U 10820 / L 30909
27A	U 30909 / L 41131	37A	U 30912 / L 30909	52A	U 41131 / L 80534
30A	U 9000 / L 30909	40A	U 30909 / L 41107	53A	U 9000 / L 70382
31A	U 30909 / L 9000	41A	U 41107 / L 30909	54A	U 9000 / L 41107
32A	U 20489 / L 50287	42A	U 9000 / L 50287	55A	U 9000 / L 10011

1953 LINCOLN

01	Regent Black	9000
02	Crown Blue Poly	10932
04	Majestic Maroon Poly	50326
05	Kingsbury Gray	30909
08	Embassy Brown Poly	20900
09	Castle Tan	20896
10	Colonial Blue	10933
12	Esquire Green Gray	30856
14	Empire Green Poly	41277
15	Oxford Gray Poly	31037
16	Palace Green Poly	41274
18	Royal Red	70592
19	Cavalier Yellow	80534

TWO-TONES

28	U 30909 / L 10932	49	U 30856 / L 41277	57	U 20900 / L 20896
29	U 10932 / L 30909	50	U 41274 / L 30856	58	U 9000
30	U 9000 / L 30909	51	U 30856 / L 41274	60	U 9000 / L 20896
31	U 30909 / L 9000	52	U 31037 / L 30909	61	U 9000 / L 50326
34	U 9000 / L 80534	53	U 30909 / L 31037	62	U 9000 / L 10933
38	U 20896 / L 50326	54	U 10932 / L 10933	63	U 9000 / L 70382
39	U 50326 / L 20896	55	U 10933 / L 10932	64	U 20896 / L 70382
48	U 41277 / L 30856	56	U 20896 / L 20900		

1954 LINCOLN

01	Regent Black	9000
02	Ambassador Blue Poly	10993
04	Majestic Maroon Poly	50326
05	Wellington Gray	31086
07	Cadet Gray Poly	31074
08	Colony Tan	20948
09	Columbia Blue	10956
11	Embassy Brown Poly	20900
11 A	Viceroy Brown Poly	21109
13	Cantebury Green	41455
14	Empire Green Poly	41277
16	Palace Green Poly	41274
19	Premier Yellow Green	41330
20	Royal Red	70592
21	Regal Red	70409
23	Ermine White	8090

TWO-TONES

34	U 9000 / L 41330	39	U 20948 / L 70592	43	U 41455 / L 41277
38	U 9000 / L 70592	39A	U 20948 / L 70409	45	U 41274 / L 41455
38A	U 9000 / L 70409	42	U 41277 / L 41455	46	U 41455 / L 41274
47	U 31074 / L 31086	56	U 9000 / L 41277	74A	U 21109 / L 8090
48	U 31086 / L 31074	57	U 9000 / L 31086	75	U 9000 / L 8090
49	U 10993 / L 10956	58	U 20948 / L 20948	76	U 8090 / L 41274
50	U 10956 / L 10993	59	U 20948 / L 10993	77	U 8090 / L 10993
51	U 20948 / L 20900	72	U 8090 / L 70592	78	U 10993 / L 8090
52	U 20900 / L 20948	72A	U 8090 / L 70409	79	U 41274 / L 8090
53	U 9000 / L 20948	73	U 8090 / L 20900	80	U 70592 / L 8090
54	U 9000 / L 10956	73A	U 8090 / L 21109	80A	U 70409 / L 8090
55	U 9000 / L 41455	74	U 20900 / L 8090		

1955 LINCOLN

01	Executive Black	9000
02	Chalet Blue Poly	11274
06	Starlight Gray	31193
07	Chancellor Gray Poly	31074
09	Brunswick Blue	10956
10	Estate Green Poly	41274
11	Viceroy Brown Poly	21109
12	Galway Green Poly	41329
13	Summit Green	41690

19 Sunstone Yellow (Green) 41330
21 Huntsman Red 70530
23 Ermine White 8090
25 Palomino Buff 21084
27 Cashmere Coral 70538
28 Taos Turquoise 41706
30 Bahama Blue Poly 11296

TWO-TONES

No	Upper	Lower	No	Upper	Lower	No	Upper	Lower
34	U 9000	L 41330	50	U 8090	L 41274	60	U 8090	L 21084
37	U 9000	L 31193	51	U 41274	L 8090	61	U 21084	L 8090
42	U 41329	L 41690	52	U 11274	L 8090	73	U 8090	L 21109
43	U 41690	L 41329	53	U 8090	L 11274	74	U 21109	L 8090
44	U 41274	L 41690	54	U 8090	L 41690	75	U 9000	L 8090
45	U 41690	L 41274	55	U 70530	L 8090	76	U 8090	L 9000
46	U 31074	L 31193	56	U 8090	L 70530	89	U 8090	L 41330
47	U 31193	L 31074	57	U 8090	L 10956	90	U 8090	L 70538
48	U 11274	L 10956	58	U 41330	L 8090	91	U 70538	L 8090
49	U 10956	L 11274	59	U 8090	L 41329	92	U 8090	L 41706
93	U 41706	L 8090	97	U 70530	L 9000	99	U 11296	L 8090
96	U 9000	L 70530	98	U 8090	L 11296			

1956 LINCOLN

01	Presidential Black	9000
02	Admiralty Blue Poly	11274
06	Centurion Gray	31306
07	Balmoral Gray Poly	31074
09	Fairmont Blue	11379
10	Shenandoah Green Poly	41800
11	Briar Brown Poly	21109
12	Wisteria	50421
13	Summit Green	41690
16	Amethyst	50393
18A	Summit Green	42048
19	Sunburst Yellow	80774
21	A Huntsman Red	70671
23	Starmist White	8050
25	Desert Buff	21084
27	Island Coral	70607
28	Taos Turquoise	41706
29	Dubonnet	50454
30	Champlain Blue	11380

TWO-TONES

No	Upper	Lower	No	Upper	Lower	No	Upper	Lower
34	U 9000	L 80774	52	U 11274	L 8050	75	U 9000	L 8050
37	U 9000	L 50393	53	U 8050	L 11274	76	U 8050	L 9000
42	U 41706	L 41690	54	U 8050	L 41690	89	U 8050	L 80774
43	U 41690	L 41706	55A	U 70671	L 8050	90	U 8050	L 50393
44	U 41800	L 41690	56A	U 8050	L 70671	91	U 8050	L 70607
45	U 41690	L 41800	57	U 8050	L 11379	92	U 8050	L 41706
46	U 31074	L 31306	58	U 11379	L 11380	93	U 8050	L 50421
47	U 31306	L 31074	59	U 11380	L 11379	96A	U 9000	L 70671
48	U 11274	L 11379	60	U 8050	L 21084	97A	U 70671	L 9000
49	U 11379	L 11274	61	U 8050	L 31306	98	U 8050	L 11380
50	U 8050	L 41800	73	U 8050	L 21109	99	U 9000	L 70607
51	U 41800	L 8050	74	U 21109	L 8050			

1956 LINCOLN CONTINENTAL

01	Black	9200
02	Deep Blue Poly	11454
04	Light Blue	11456
05	Deep Green Poly	41892
07	Light Green	41894
08	Deep Bronze Poly	21226
10	Beige	21228
11	Deep Gray Poly	31332
13	Deep Red	50429
14	White	8112
15	Medium Blue Poly	11455
16	Medium Green Poly	41893
17	Medium Bronze Poly	21227
18	Medium Gray Poly	31333

TWO-TONES

No	Upper	No	Lower	No	Upper	No	Lower	No	Upper	No	Lower
02	U 9200	15	L 11455	08	U 21226	17	L 21227	11	U 31332	14	L 8112
05	U 41892	16	L 41893	11	U 31332	18	L 31333				

1957 LINCOLN

No.	Color	Code
01	Presidential Black	9000
03	Gainsborough Blue Poly	11558
04	Seascape Blue	11555
05	Horizon Blue	11567
06	Starmist White	8050
07	Ivy Green Poly	42045
09	Willow Green (Blue)	11578
10	Sand	21311
11	Desert Buff	21084
12	Huntsman Red	70671
13	Saturn Gold	80874
14	Taos Turquoise	41706
15/29	Dubonnet Poly	50454
16	Cinnamon Poly	21314
17	Bermuda Coral	70728
18	Flamingo	70727
21	Oxford Gray Poly	31412
22	Vermont Green Poly	42038

TWO-TONES

No.		No.		No.	
38	U 11567	56	U 21084	76	U 80874
	L 11558		L 21311		L 9000
39	U 11558	57	U 8050	77	U 8050
	L 11567		L 21084		L 80874
40	U 11555	58	U 21084	78	U 80874
	L 11558		L 8050		L 8050
41	U 11558	59	U 9000	79	U 8050
	L 11555		L 21084		L 9000
42	U 8050	60	U 11578	80	U 9000
	L 11558		L 42045		L 8050
43	U 11558	61	U 42045	81	U 9000
	L 8050		L 11578		L 70671
44	U 11567	62	U 42038	82	U 70671
	L 11555		L 42045		L 9000
45	U 11555	63	U 42045	83	U 31412
	L 11567		L 42038		L 70671
46	U 8050	64	U 8050	84	U 21311
	L 11555		L 42045		L 70671
47	U 11555	65	U 42045	85	U 8050
	L 8050		L 8050		L 70671
48	U 8050	66	U 11578	86	U 70671
	L 11567		L 42038		L 8050
49	U 21311	67	U 42038	87	U 9000
	L 11558		L 11578		L 41706
50	U 9000	70	U 8050	88	U 41706
	L 11567		L 11578		L 9000
51	U 8050	71	U 11578	89	U 8050
	L 21311		L 8050		L 41706
52	U 21311	72	U 21311	90	U 41706
	L 8050		L 42045		L 8050

No.		No.		No.	
53	U 9000	73	U 21311	91	U 9000
	L 21311		L 42038		L 50454
54	U 21311	74	U 9000	92	U 8050
	L 9000		L 11578		L 50454
55	U 21311	75	U 9000	93	U 50454
	L 21084		L 80874		L 8050
94	U 21311	102	U 8050	110	U 50454
	L 50454		L 70728		L 70727
95	U 9000	103	U 70728	111	U 70727
	L 21314		L 8050		L 50454
96	U 21311	104	U 9000	136	U 8050
	L 21314		L 70727		L 31412
97	U 21314	105	U 70727	137	U 31412
	L 21311		L 9000		L 8050
98	U 8050	106	U 8050	146	U 8050
	L 21314		L 70727		L 42038
99	U 21314	107	U 70727	147	U 42038
	L 8050		L 8050		L 8050
100	U 9000	108	U 31412		
	L 70728		L 70727		
101	U 70728	109	U 70727		
	L 9000		L 31412		

In the two-tone combinations above the letter (U) indicates Upper Body, (L) Lower Body.

1957 LINCOLN CONTINENTAL

No.	Color	Code
01	Black	9200
02	Deep Blue Poly	11454
04	Light Blue	11456
05	Deep Green Poly	41892
07	Light Green	41894
08	Deep Bronze Poly	21226
10	Beige	21228
11	Deep Gray Poly	31332
13	Deep Red	50429
14	White	8112
15	Medium Blue Poly	11455
16	Medium Green Poly	41893
17	Medium Bronze Poly	21227
18	Medium Gray Poly	31333
19	Light Blue Poly	11670
20	Light Green Poly	42140
21	Tan Poly	21448
22	Silver Poly	31553

1958 LINCOLN

No.	Color	Code
01	Presidential Black	9000
02	Spartan Gray Poly	31455
03	Athenian Gray	31459
04	Arrowhead Blue Poly	11680
05	Seneca Blue Poly	11661

Code	Color	Paint
06	Shasta Blue	11662
07	Starmist White	8050
08	Spruce Green Poly	42148
10	Sequoia Green	42086
11	Jade	42093
12	Suede	21374
13	Deauville Yellow	80918
15	Autumn Rose	70789
16	Sunset	70775
18	Matador Red	70781
19	Claret Poly	50482
20	Champagne	21460
23	Platinum	11683
24	Silver Poly	31608
25	Copper Poly	21517
26	Rosemetal Poly	31656

TWO-TONE COMBINATIONS

42,43	31459 31455	83,84	8050 21374	159,160	80918 31455
44,45	8050 31455	87,88	8050 42093	163,164	21460 42086
46,47	8050 31459	89,90	8050 70775	187,188	9000 11683
48,49	11662 11680	95,96	8050 70789	189,190	11680 11683
50,51	8050 11680	99,100	8050 50482	191,192	11661 11683
52,53	11662 11661	101,102	70775 31455	193,194	11662 11683
54,55	8050 11661	103,104	9000 21460	195,196	31455 11683
56	8050 11662	105,106	8050 21460	203,204	31608 9000
57,58	8050 9000	115	8050 11662	205,206	31608 8050
59,60	42086 42148	116	8050 42148	207,208	31608 80918
61	8050 42148	117	8050 80918	227,228	31608 70789
68,69	42093 42086	126,127	21460 42093	229,230	31680 42093
70,71	8050 42086	132,133	21460 42148	233,234	31608 50482
72,73	9000 80918	137	31459 70775	237,238	31608 70775
74	8050 80918	139	21460 70781	243,244	31608 11662
79,80	9000 70871	143,144	21460 21374		

81,82 8050 145,146 80918
70871 21374

1959 LINCOLN

Code	Color	Paint
01	Presidential Black	9000
07	Glacier White	8103
15	Linden Green	42396
16	Palm Green	42408
17	Peacock Green Poly	42399
27	Crystal Blue	11990
28	Pearl Blue Poly	11757
29	Midnight Blue Poly	11930
40	Claret Poly	50482
45	Bolero Red	70869
50	Warwick Gray	31767
55	Silver Poly	31608
60	Tawn	21400
66	Sunstone	21527
75	Deauville Yellow	80918
80	Platinum	11683
81	Copper Poly	21517
	Early Production	
82	Copper Poly	21473
83	Cameo Rose	70883
88	Aquamarine	11920
92	Burnished Gold Poly	21648
97	Sapphire Poly	11921

NOTE: On two-tone combinations the first two paint code numbers indicate the upper color, the second two numbers indicate the lower color.

1960 LINCOLN

Code	Color	Paint
A	Presidential Black	9000
B	Marine Blue Poly	12144
C	Pale Turquoise	42434
D	Sapphire Poly	11921
E	Electric Blue Poly	12164
F	Blue Crystal	12147
G	Tawney Beige	21794
I	Pale Turquoise	12146
J	Cherokee Red	71054
L	Gold Dust Poly	21886
M	Polaris White	8238
N	Platinum	11683
Q	Copper Poly	21473
R	Pastel Yellow	81052
S	Deerfield Green Poly	42663
T	Terra Verde Green Poly	42344
U	Metallic Rose Glow Poly	21849
V	Twilight Pink	71055
W	Killarney Green	42634

X	Maple Leaf Poly	50538
Y	Spartan Gray Poly	31956
Z	Cloud Silver Poly	31991

In two-tone combinations the first letter indicates the lower color, the second letter upper.

1961 LINCOLN CONTINENTAL

A	Presidential Black	9000
B	Royal Red Poly	71110
C	Turquoise Mist	42434
D	Blue Haze	12355
E	Saxon Green Poly	42820
F	Sunburst Yellow	81249
H	Empress Blue Poly	12366
I	Green Velvet Poly	42805
K	Crystal Green Poly	42825
M	Sultana White	8238
N	Platinum	11683
P	Executive Gray Poly	32081
Q	Sheffield Gray Poly	32089
R	Columbia Blue Haze Poly	12361
T	Honey Beige	81224
U	Rose Glow Poly	21849
V	Summer Rose	71055
W	Regency Turquoise Poly	12359
X	Black Cherry Poly	50593
Y	Briar Brown Poly	21959
Z	Desert Frost Poly	21958

In two-tone combinations the first letter indicates the lower color, the second letter upper.

1962 LINCOLN

A	Presidential Black	9000
B	Royal Red Poly	71110
C	Oxford Gray Poly	32156
D	Riviera Turquoise Poly	11921
E	Bermuda Blue Poly	12164
F	Powder Blue	12147
G	Silver Mink Poly	12497
H	Nocturne Blue Poly	12547
I	Castilian Gold Poly	31842
L	Teaberry	71239
M	Sultana White	8238
N	Platinum	11683
P	Scotch Green Poly	42929
R	Jamaica Yellow	81324
S	Highlander Green Poly	42925
T	Champagne	22110
U	Velvet Turquoise Poly	12493
V	Chestnut Poly	60392
X	Black Cherry Poly	50593

| Z | Desert Frost Poly | 21958 |

In two-tone combinatiodns the first letter indicates the lower color, the second letter upper.

1963 LINCOLN CONTINENTAL

A	Black Satin	9000
C	Oxford Gray Poly	32156
D	Riviera Turquoise Poly	11921
E	Bermuda Blue Poly	12164
G	Silver Mink Poly	12497
H	Nocturne Blue Poly	12547
I	Polynesian Gold Poly	32277
L	Teaberry	71239
M	Ermine White	8238
N	Platinum	11683
O	Inverness Green Poly	43082
Q	Spanish Red Poly	71327
R	Premier Yellow	81324
S	Highlander Green Poly	42925
T	Nassau Beige	22110
W	Rose Poly	71322
X	Burgundy Frost Poly	50593
Z	Autumn Frost Poly	21958

In two-tone combinations the first letter indicates the lower color, the second letter upper.

1964 LINCOLN-CONTINENTAL & FORD THUNDERBIRD

A LC	Black Satin	9000
A TB	Raven Black	
B TB	Pagoda Green	12851
C LC	Princeton Gray Poly	32251
C TB	Gunmetal Gray Poly	
E LC	Silvermink Poly	12751
E TB	Silver Blue Poly	
F LC	Arcadian Blue	12854
F TB	Powder Blue	
G LC	Buckskin Tan	22230
G TB	Prairie Tan	
H LC	Caspian Blue Poly	12752
H TB	Nocturne Blue Poly	
I TB	Florentine Green	22395
J LC	Rangoon Red	71243
J TB	Fiesta Red	
L TB	Samoan Coral	60398
M LC	Wimbledon White	8378
M TB	Artic White	
N LC	Platinum Blue	11683
N TB	Diamond Blue	
O LC	Silver Green Poly	43230
P LC	Burnished Bronze Poly	22475
P TB	Prairie Bronze Poly	

Q	Brittany Blue Poly	12843
Q	Huron Blue Poly	
R LC	Phoenician Yellow	81444
R TB	Encino Yellow	
S LC	Cascade Green Poly	43148
S TB	Highlander Green Poly	
T LC	Desert Sand	22249
T TB	Navajo Beige	
U LC	Patrician Green Poly	12745
U TB	Regal Turquoise Poly	
V	Sunlight Yellow	81467
W LC,TB	Rose Beige Poly	71358
X LC	Vintage Burgundy Poly	50669
X TB	Royal Maroon Poly	
Z LC	Chantilly Beige Poly	22424
Z TB	Silver Sand Poly	

TB - Thunderbird, LC - Lincoln-Continental
In two-tone combinations the first letter indicates lower color, the second letter upper.

1965 LINCOLN-CONTINENTAL & FORD THUNDERBIRD

A LC	Black Satin	9000,9300
A TB	Raven Black	
B LC	Turino Turquoise Poly	12748
B TB	Midnight Turquoise Poly	
C LC	Persian Gold Poly	22439
C TB	Honey Gold Poly	
E LC	Madison Gray Poly	12751
E TB	Silver Mink Poly	
F LC	Powder Blue	12854
F	Arcadian Blue	
G LC	Willow Gold	81460
G TB	Pastel Yellow	
H LC	Nocturne Blue Poly	12752
H TB	Caspian Blue Poly	
J LC	Fiesta Red	71243
J TB	Rangoon Red	
L LC	Heather Poly	50670
M LC	Arctic White	8378
M TB	Wimbledon White	
N LC	Platinum	11683
N TB	Diamond Blue	
P LC	Burnished Bronze Poly	22475
P TB	Prairie Bronze Poly	
Q LC	Huron Blue Poly	12843
Q TB	Brittany Blue Poly	
R LC	Spanish Moss Poly	43337
R TB	Ivy Green Poly	
S LC	Charcoal Frost Poly	32390
S TB	Charcoal Gray Poly	
T LC	Desert Sand	22249

T TB	Navajo Beige	
U LC	Teal Poly	12745
U TB	Patrician Green Poly	
V LC	Russet Poly	22610
V TB	Emberglo Poly	
W TB	Rose Beige Poly	71358
X LC	Royal Maroon Poly	50669
X TB	Vintage Burgundy Poly	
Z LC	Silver Sand Poly	22424
Z TB	Chantilly Beige Poly	
4 LC	Neptune Blue	12876
4 TB	Frost Turquoise	
	Silver Gray	32500
	(used on Lincoln Stone Shield)	

LC - Lincoln-Continental, TB - Thunderbird
In two-tone combinations the first letter indicates lower color, the second letter upper.

1966 LINCOLN-CONTINENTAL & FORD THUNDERBIRD

A LC	Black Satin	9000,9300
A TB	Raven Black	
B LC	Chesterfield Beige Poly	22609
B TB	Sundust Beige	
E LC	Madison Gray Poly	12751
E TB	Silver Mink Poly	
F LC	Powder Blue	12854
F TB	Arcadian Blue	
G TB	Sapphire Blue Poly	13075
H LC	Teakwood	22528
H TB	Sahara Beige	
K LC	Pitcairn Blue Poly	13076
K TB	Nightmist Blue Poly	
L LC	Venetian Yellow	43354
L TB	Honeydew Yellow	
M LC	Arctic White	8378
M TB	Wimbledon White	
N LC	Platinum	11683
N TB	Diamond Blue	
O LC	Rose Mist Poly	22615
O TB	Silver Rose Poly	
P LC	Sandalwood Poly	22603
P TB	Antique Bronze Poly	
Q LC	Huron Blue Poly	12843
Q TB	Brittany Blue Poly	
R LC	Spanish Moss Poly	43408
R TB	Ivy Green Poly	
S LC	Charcoal Frost Poly	32390
T LC	Cranberry	71528
T TB	Candyapple Red	
U LC	Teal Poly	12745
U TB	Tahoe Turquoise Poly	

V LC	Russet Poly	22610
V TB	Emberglo Poly	
X LC	Royal Maroon Poly	50669
X TB	Vintage Burgundy Poly	
Z LC	Florentine Gold Poly	43433
Z TB	Sauterne Gold Poly	
2 LC	Empress Turquoise Poly	43454
2 TB	Mariner Turquoise Poly	

LC - Lincoln-Continental, TB - Thunderbird
In two-tone combinations the first letter or digit indicates lower color, the second letter upper.

1967 LINCOLN-CONTINENTAL & FORD THUNDERBIRD

A LC	Black Satin	9000,9300
A TB	Raven Black	
B LC	Palomar Blue	12876
B TB	Frost Turquoise	
C LC	Huntington Gray Poly	32485
C TB	Charcoal Gray Poly	
E LC	Antique Beige Poly	22711
E TB	Beige Mist Poly	
F LC	Powder Blue	12854
F TB	Arcadian Blue	
H LC	Cameo Green	43575
H TB	Diamond Green	
K LC	Pitcairn Blue Poly	13076
K TB	Nightmist Blue Poly	
M LC	Arctic White	8378
M TB	Wimbledon White	
M LC	Arctic White #2	8734
M TB	Wimbledon White	
N LC	Platinum	11683
N TB	Diamond Blue	
P LC	Champagne Poly	22744
P TB	Pewter Mist Poly	
Q LC	Huron Blue Poly	12843
Q TB	Brittany Blue Poly	
R LC	Spanish Moss Poly	43408
R TB	Ivy Green Poly	
T LC	Cranberry	71528
T TB	Candyapple Red	
U LC	Teal Poly	12745
U	Tahoe Turquoise Poly	
V LC	Aegean Bronze Poly	22749
V TB	Burnt Amber Poly	
X LC	Royal Maroon Poly	50669
X TB	Vintage Burgundy Poly	
Z LC	Florentine Gold Poly	43433
Z TB	Sauterne Gold Poly	
2 LC	Granada Yellow	81444
2 TB	Phoenician Yellow	

4 LC	Silver Mist Poly	32520
4 TB	Silver Frost Poly	
6 LC	Desert Sand	22249
6 TB	Pebble Beige	

LC - Lincoln-Continental, TB - Thunderbird
In two-tone combinations the first letter or digit indicates lower color, the second letter upper.

1968 LINCOLN-CONTINENTAL & FORD THUNDERBIRD

A LC	Black Satin	9000,9300
A TB	Raven Black	
B LC	Royal Burgundy	50746
B TB	Royal Maroon	
E LC	Antique Beige Poly	22711
E TB	Beige Mist Poly	43645
G LC	Belmont Green Poly	
H LC	Cameo Green	43575
H TB	Diamond Green	
I LC	Aspen Green Poly	43576
I TB	Lime Gold Poly	
J LC	Mediterranean Poly	13342
J TB	Midnight Aqua Poly	
L LC	Foxcroft Silver Poly	32608
L TB	Silver Pearl Poly	
M LC	Arctic White	8378
M TB	Wimbledon White	
M LC	Arctic White #2	8734
M TB	Wimbledon White #2	
N LC	Platinum	11683
N TB	Diamond Blue	
P LC	Champagne Poly	22744
P TB	Pewter Mist Poly	
Q LC	Huron Blue Poly	13619
Q TB	Brittany Blue Poly	
R LC	Grenoble Green Poly	43644
R TB	Highland Green Poly	
S LC	Ascot Gray Poly	50670
T LC	Cranberry	71528
T TB	Candyapple Red	
U LC	Teal Poly	12745
U TB	Tahoe Turquoise Poly	
V LC	Daulton Blue	13317
V TB	Alaska Blue	
W LC	Mikado Yellow	81584
W TB	Meadowlark Yellow	
X LC	Admiralty Blue Poly	13356
X TB	Presidential Blue Poly	
Y LC	Chancery Gold Poly	22833
Y TB	Sunlit Gold Poly	
Z LC	Eton Gray Poly	32527
Z TB	Oxford Gray Poly	

| 6 LC | Desert Sand | 22249 |
| 6 TB | Pebble Beige | |

LC - Lincoln-Continental, TB - Thunderbird

1969 LINCOLN-CONTINENTAL, MARK III AND FORD THUNDERBIRD

B L,M	Maroon	50746
T	Royal Maroon	
C L,M	Dark Ivy Green Poly	2037
T	Black Jade Poly	
G L,M	Medium Orchid Poly	2040
T	Lilac Frost Poly	
H L,M	Light Green	43575
T	Diamond Green	
I L,M	Medium Lime Poly	2054
T	Lime Gold Poly	
J L,M	Dark Aqua Poly	13342
T	Midnight Aqua Poly	
K L,M	Dark Orchid Poly	2041
T	Midnight Orchid Poly	
L L,M	Light Gray Poly	32608
M L,M	White	8378, 8723#2
T	Wimbleden White	
N L,M	Platinum	11683
T	Diamond Blue	
O L,M	Medium Green Poly	43645
Q L,M	Medium Blue Poly	13619
T	Brittany Blue Poly	
R L,M	Light Gold	2043
T	Morning Gold	
S L,M	Medium Gold Poly	2044
T	Champagne Gold Poly	
T L,M	Red	71528
T	Candyapple Red	
U L,M	Medium Aqua Poly	12745
T	Tahoe Turquoise Poly	
V L,M	Light Copper Poly	2045
T	Copper Flame Poly	
W L,M	Yellow	81584
T	Meadowlark Yellow	
X L,M	Dark Blue Poly	2056
T	Presidential Blue Poly	
Y L,M	Burnt Orange Poly	2046
T	Indian Fire Poly	
Z L,M	Dark Gray Poly	32527
T	Oxford Gray Poly	
A L,M	Black	9000,9300
T	Raven Black	

L-Lincoln, M-Mark, T-Thunderbird

1970 LINCOLN-CONTINENTAL, MARK III AND FORD THUNDERBIRD

B	Dark Maroon	2150
C L,M	Dark Green Poly	2146
T	Dk. Ivy Green Poly	
E	Light Blue	12997
F	Dk. Bright Aqua Poly	13329
H L,M	Light Green	43575
I L,M	Light Green	43575
A L,M	Black	9000,9300
T	Raven Black	
J	Deep Blue Poly	2137
L	Light Gray	2144
M L,M	White	8378
T	Wimbledon White	
O	Medium Green Poly	43645
P	Med. Ivy Green Poly	2147
Q	Medium Blue Poly	2138
R	Dark Brown Poly	2140
S	Medium Gold Poly	2044
T T	Candyapple Red	71528
U L,M	Medium Aqua Poly	12745
X	Dark Blue	2139
Y	Medium Bronze Poly	2142
Z	Dark Gray Poly	2145
2 L,M	Light Ivy Yellow	2047
T	New Lime	
5	Medium Brown Poly	2160
8 L,M	Light Gold	2043
T	Morning Gold	
9	Pastel Yellow	81584
19 L,M	Green Stardust Poly	2239
T	Green Fire Poly	
59 L,M	Red Stardust Poly	2158
T	Burgundy Fire Poly	
89 L,M	Bronze Stardust Poly	2143, 2238
T	Bronze Fire Poly	

M - Mark III, L - Lincoln-Continental, T - Thunderbird
In two-tone combinations the first letter or digit indicates lower color; the second letter or digit indicates upper color.

1971 LINCOLN-CONTINENTAL, MARK III AND FORD THUNDERBIRD

B	Maroon Poly	2295
F	Bright Aqua Poly	2282
G	Dark Green	2288
H	Light Green	2290
J-9	Anniversary Gold Poly	2389
K	Dark Gray Poly	2145
L	L,M Light Gray Poly	2144

M	L,M White	8378
T	Wimbledon White	
N	Pastel Blue	11683
O	Light Yellow Gold	2289
Q	Medium Blue Poly	2372
R	Dark Brown Poly	2140
S	Gray Gold Poly	2326
T	L,M Red	71528
T	Candyapple Red	
V	Light Pewter Poly	2287
W	Yellow	2157
X	Dark Blue Poly	13076
Y	Deep Blue Poly	2137
2	Tan	2285
5	Medium Brown Poly	2371
6	Bright Blue Poly	13357
7	Maroon	50746
39	L,M Ginger Bronze Moondust Poly	2286
T	Walnut Fire Poly	
C-9 L,M	Red Moondust Poly	2361
T	Burgundy Fire Poly	
D-9 L,M	Blue Moondust Poly	2283
T	Blue Fire Poly	
E-9 L,M	Ivory Bronze Moondust Poly	2360
T	Green Fire Poly	
A	L,M Black	9000,9300
T	Raven Black	

M - Mark III, L - Lincoln-Continental, T - Thunderbird

1972 LINCOLN-CONTINENTAL, MARK IV AND FORD THUNDERBIRD

1-A	Light Gray Poly	2409
1-D	M Moondust Poly	Silver 2410
2-G	L,M Red Moondust Poly	2361
T	Burgundy Fire Poly	
2-J,7	Maroon	50746
3-B	Light Blue	2403
3-C	L,M Blue Moondust Poly	2283
T	Blue Fire Poly	
3-D	Medium Blue Poly	2404
3-H	Dark Blue Poly	2405
1-C	Black	9000,9300
4-A	Pastel Lime	2411
4-B	Bright Green Gold Poly	2406
4-D	L,M Ivy Moondust Poly	2360
T	Green Fire Poly	
4-G	L,M Light Ivy Moondust Poly	2475
T	Lime Fire Poly	
4-P	Medium Green Poly	2289

4-Q	Dark Green Poly	2291
4-U	L Lime Gold Moondust Poly	2500
T	Emerald Fire Poly	
5-C	L,M Ginger Moondust Poly	2286
T	Walnut Fire Poly	
5-D	L,M Light Ginger Moondust Poly	2474
T	Cinnamon Fire Poly	
5-F	Dark Brown Poly	2407
5-G	L,M Copper Moondust Poly	2408
T	Copper Fire Poly	
6-B	Light Yellow Gold	2298
6-D	Yellow	2157
6-G	L,M Gold Moondust Poly	2389
T	Gold Fire Poly	
6-J	Gray Gold Poly	2326
9-A	White	8378

L-Lincoln, M-Mark, T-Thunderbird

1973 LINCOLN-CONTINENTAL

1-A	Light Gray Poly	2409
1-C	Black	9000,9300
1-D	Silver Moondust Poly	2410
2-G L,MK	Red Moondust Poly	2361
T	Burgundy Fire Poly	
2-J	Maroon	50746
3-A	Pastel Blue	11683
3-D	Medium Blue Poly	2404
3-G	Dark Blue Poly	2472
3-L L,MK	Silver Blue Moondust Poly	2503
T	Silver Blue Fire Poly	2404
4-D L,MK	Ivy Moondust Poly	2360
T	Green Fire Poly	
4-P	Medium Green Poly	2289
4-Q	Dark Green Poly	2291
4-S	Light Green	2419
4-U	L,MK Brt.Lime Gold Moondust Poly	2500
T	Emerald Fire Poly	
4-Y	Green Diamond	2596
5-D L,MK	Lt. Ginger Moondust Poly	2474
T	Cinnamon Fire Poly	
5-F	Dark Brown Poly	2407
5-K L,MK	Ginger Moondust Poly	2505
T	Almond Fire Poly	
5-L	Tan	2285
5-P L,MK	Dk. Copper Moondust Poly	2498
T	Mahogany Fire Poly	
5-Q	Dark Brown Poly	2616
5-R T	Gold Poly	2592
5-S	Medium Beige	2617
6-B	Light Yellow Gold	2298

6-D	Yellow	2157
6-G L,MK	Gold Moondust Poly	2389
T	Gold Fire Poly	
6-L	Medium Gold Poly	2497
9-C	White Decor	2512
9-A	White	8378
52	Copper Diamond Flare Poly	2594

L-Lincoln, M-Mark, T-Thunderbird

In two-tone combinations the first two digits indicate lower color; the second two digits indicate upper color.

1974 LINCOLN CONTINENTAL

1-C	Black	9000,9300
1-D	Silver Moondust Poly	2410
1-E	Light Silver Cloud Poly	2608
2-G L,MK	Red Moondust Poly	2361
T	Burgundy Fire Poly	
3-A	Platinum	11683
3-D	Medium Blue Poly	2404
3-G	Dark Blue Poly	2472
3-L L,MK	Silver Blue Moondust Poly	2503
T	Silver Blue Fire Poly	
3-P	Silver Blue Diamond Flare Poly	2595
4-A	Pastel Lime	2411
4-U	Lime Gold Moondust Poly	2500
4-Y	Green Diamond Flare Poly	2596
5-Q	Dark Brown Poly	2616
5-R	Bright Gold Bronze Poly	2592
5-S	Medium Beige	2617
5-V	Buff	2619
5-Z	Gold Poly	2667
6-G L,MK	Yellow Gold Moondust Poly	2389
T	Gold Fire Poly	
6-L	Medium Gold Poly	2497
6-M	Medium Dark Gold Poly	2621
6-N	Maize Yellow	2622
9-C	White Decor	2512
9-D	White	2684
51	Ginger Diamond Flare Poly	2655
52	Copper Diamond Flare Poly	2594
54	MK Unique Gold Diamond Fire Poly	2682

L-Lincoln, M-Mark, T-Thunderbird

In two-tone combinations the first two digits indicate lower color; the second two digits indicate upper color.

1975 LINCOLN CONTINENTAL

1-C	Black	9000,9300
1-J L,MK	Silver Diamond Fire Poly	2715
T	Silver Starfire Poly	
2-G L,MK	Red Moondust Poly	2361
T	Burgundy Fire Poly	
2-M	Dark Red	2609
2-P L	Medium Taupe Diamond Fire Poly	2718
3-G	Dark Blue Poly	2472
3-P L,MK	Blue Diamond Fire Poly	2595
T	Blue Starfire Poly	
3-Q L,MK	Medium Pastel Blue	2613
T	Pastel Blue	
3-R L,MK	Silver Blue Diamond Fire Poly	2719
T	Silver Blue Starfire Poly	
4-U	Lime Gold Moondust Poly	2500
4-V	Dark Yellow Green Poly	2614
4-Z	Light Gold Poly	2720
5-Q	Dark Brown Poly	2616
6-G L,MK	Yellow Gold Moondust Poly	2389
T	Gold Fire Poly	2720
6-N	Medium Ivy Yellow	2622
6-P	Cream	2790
6-Q	Dark Gold Poly	2721
7-F	Medium Jade Diamond Flare Poly	2809
9-D	White	2684
41	L,MK Bright Lime Gold Diamond Fire Poly	2722
T	Bright Lime Starfire Poly	
45	MK Aqua Blue Deamond Fire Poly	2724
46	L Dark Jade Poly	2725
47	Light Green	2726
51	L,MK Ginger Bronze Diamond Fire Poly	2655
T	Ginger Bronze Starfire Poly	
52	L,MK Copper Diamond Fire Poly	2594
T	Copper Starfire Poly	
54	L,MK Unique Gold Diamond Fire Poly	2682
T	Unique Gold Starfire Poly	

MK - Mark IV, L - Lincoln-Continental, T - Thunderbird

In two-tone combinations the first two digits indicate lower color, the second two digits indicate upper color.

1976 LINCOLN CONTINENTAL

1-C	Black	9000,9300
1-J L,MK	Silver Diad. Fire Poly	2715
T	Silver Starfire Poly	
1-L L,MK	Black Diamond Flare Poly	2882
1-N	Dove Gray	2847
2-G L,MK	Red Moondust Poly	2361

T	Burgundy Fire Poly	
2-M	Dark Red	2609
2-P L	Medium Taupe Diad Fire Poly	2718
2-S	Dark Red Poly	2831
2-T	MK Rose Crystal Poly	2832
2-U	MK Lipstick Red	2833
3-G	Dark Blue Poly	2472
3-P L,MK	Blue Diad. Fire Poly	2595
T	Blue Starfire Poly	
3-S	Light Blue	2834
5-Q	Dark Brown Poly	2616
6-P	Cream	2790
6-U	Tan	2836
6-Y	L,MK Brt. Yl. Gd. Diad. Fire Poly	2839
T	Brt. Yl. Gd. Starfire Poly	
7-A	Light Jade	2840
7-B	MK Light Jade Crystal Poly	2841
7-F	T Med. Jade Starfire Poly	2809
8-A	MK Light Apricot Crystal Poly	2842
9-D	White	2684
45	MK Aqua Blue Diad. Fire Poly	2724
46	Dark Jade Poly	2725
51	L,MK Gin. Brnze. Diad. Fire Poly	2655
T	Gin. Brnze. Starfire Poly	
54	L Unique Gold Diad. Fire Poly	2684
59	L,MK Med. Chestnut Diad. FirePoly	2843

L-Lincoln, M-Mark, T-Thunderbird
In two-tone combinations the first two digits indicate lower color; the second two digits indicate upper color.

1977 LINCOLN

1-C	Black	9000,9300
1-J	Sil. Diad. Fire Poly	2715
1-L	Blk. Diad. Fire Poly	2882
1-N	Dove Gray	2847
1-P	Med. Gray Poly (Two-Tone)	2967
2-G	Red Moondust Poly	2361
3-G	Brt. Drk. Blue Poly	2472
2-S	Drk. Diad. Brt. Red Poly	2831
2-T	Rose Crystal Poly	2832
3-U	Light Blue	2907
6-P	Cream	2790
6-Y	Brt. Yel. Gold Diad. Fire Poly	2839
7-B	Light Jade Crystal Poly	2841
8-N	Drk. Cordovan Poly	2920
8-P	Light Cordovan	2921
8-V	Med. Emb. Diad. Brt. Poly	2922
8-Z	Med. Nect. Diad. Brt. Poly	2919

9-D	White	2684
31	Midnight Blue	2925
32	Med. Blue Diad. Brt. Poly	2909
46	Dark Jade Poly	2725
84	Midnight Cordovan	2990

In two-tone combinations the first two digits indicate lower color; the second two digits indicate upper color.

1978 LINCOLN

1-C	Black	9000,9300
1-S	Med. Silver Poly	2930
1-W	Black	3037
1-Y	Light Silver Poly	3036
2-G	Med. Red Poly	2361
5-A	Drk. Champagne Poly	3038
5-C	Champagne Poly	3039
5-D	Light Champagne	3040
5-G	Light Chamois	2989
5-K	Med. Chamois	2997
5-L	Midnight Cordovan	2995
5-R	Drk. Cordovan Poly	2950
6-P	Cream	2790
7-V	Midnight Jade	3043
7-Y	Light Jade Poly	3045
9-D	White	2684
9-F	White	2996
23	Dark Red Poly	3047
31	Midnight Blue	2925
32	Med. Bl. Diad. Flare Poly	2909
34	Medium Blue	3048
35	Midnight Blue Poly	2952
36	Wedgewood Blue	2994
37	Med. Blue Poly	3049
38	Diad. Blue Poly	3050
52	Light Champagne	24416
66	Jubilee Gold Poly	3054
67	Gold Poly	3055
68	Cream	3052
72	Midnight Jade	3042
83	Light Chamois	3057
84	Midnight Cordovan	2990
88	Crystal Apricot Poly	3059

In two-tone combinations the first two digits indicate lower color; the second two digits indicate upper color.

PAINT CODE	COLOR	PPG CODE
1949 MERCURY		

See 1949 Mercury Colors (Model Code "M") listed under 1949 Lincoln.

PAINT CODE	COLOR	PPG CODE
1950 MERCURY		
	Black	9000
	Banning Blue Poly	10582
	Laguana Blue Poly	10659
	Dune Beige	20658
	Penrod Tan	20659
	Trojan Gray	30717
	Maywood Green Poly	40874
	Everglade Green	40894
	Roanoke Green Poly	40860
	Royal Bronze Maroon Poly	50126
	Mirada Yellow	80480

TWO-TONES

U 40874	U 40860
L 30717	L 30717
U 20659	U 10659
L 20658	L 30717
U 30717	U 10582
L 10582	L 30717
U 30717	
L 40860	
U 30717	
L 10659	

PAINT CODE	COLOR	PPG CODE
1951 MERCURY		
01	Black	9000
03	Banning Blue Poly	10582
05	Luxor Maroon Poly	50218
07	Mission Gray	30857
09	Tomah Ivory	80526
11	Everglade Green	40894
13	Sheffield Green	40944
15	Kerry Blue Poly	10764
15C	Academy Blue	10011
17	Coventry Green Gray	30856
19	Vassar Yellow	80534
202	Monterey Red Poly	50224
203	Turquoise Blue	10761
204	Brewster Green Poly	41088
204	Yosemite Green Poly	40180

TWO-TONES

30	U 9000	33	U 30856	38	U 30856
	L 80526		L 40894		L 40944
31	U 80526	34	U 40894	39	U 30857
	L 40944		L 30856		L 10582
32	U 40944	37	U 40944	40	U 10582
	L 80526		L 30856		L 30857

TWO-TONES MONTEREY MODEL

202	U Black (Cloth)	204	U Oyster White (Cloth)
	L 41088		L 50224
203	U Seal Brown (Cloth)	204	U Oyster White (Cloth)
	L 10761		L 40180

PAINT CODE	COLOR	PPG CODE
1952 MERCURY		
01A	Black	9000
02A	Admiral Blue	10820
04A	Fanfare Maroon	50287
05A	Newport Gray	30909
07A	Lucerne Blue	10797
09A	Pebble Tan	20489
10A	Academy Blue	10011
11A	Hillcrest Green	41131
12A	Coventry Green Gray	30856
13A	Lakewood Green	41107
19A	Vassar Yellow	80534

TWO-TONES

20A	U 10797	30A	U 9000	42A	U 9000
	L 10820		L 30909		L 50287
21A	U 10820	31A	U 30909	43A	U 9000
	L 10797		L 9000		L 30856
24A	U 41131	32A	U 20489	46A	U 9000
	L 30856		L 50287		L 20489
25A	U 30856	33A	U 50287	47A	U 9000
	L 41131		L 20489		L 10797
28A	U 41107	34A	U 9000	54A	U 9000
	L 30856		L 80534		L 41107
29A	U 30856	35A	U 80534		
	L 41107		L 900		

PAINT CODE	COLOR	PPG CODE
1953 MERCURY		
01	Mercury Black	9000
02	Superior Blue Poly	10932
04	Mohawk Maroon Poly	50326
05	Glenwood Gray	30909
06	Beechwood Brown Poly	20899
06A	Brentwood Brown Poly	20944

07	Banff Blue	10927
09	Tahiti Tan	20896
11	Sherwood Green Poly	41279
12	Pinehurst Green Gray	30856
12A	Asheville Green	41378
13	Village Green Poly	41280
17	Bittersweet	60135
19	Yosemite Yellow	80534
20	Siren Red	70409

TWO-TONES

26	U 10927 / L 10932	37	U 30856 / L 41820	58	U 9000 / L 30856
27	U 10932 / L 10927	37A	U 41378 / L 41280	59	U 9000 / L 41280
28	U 30909 / L 10932	38	U 20896 / L 50326	60	U 9000 / L 20896
29	U 10932 / L 30909	39	U 50326 / L 20896	61	U 9000 / L 50326
30	U 9000 / L 30909	40	U 20896 / L 20899	66	U 70409 / L 9000
31	U 30909 / L 9000	40A	U 20896 / L 20900	67	U 41280 / L 20896
32	U 41279 / L 30856	41	U 20899 / L 20896	68	U 20896 / L 41280
32A	U 41279 / L 41378	41A	U 20900 / L 20896	69	U 41279 / L 80534
33	U 30856 / L 41279	42	U 20896 / L 60135	70	U 80534 / L 41279
33A	U 41378 / L 41279	43	U 60135 / L 20896	71	U 20900 / L 41378
34	U 9000 / L 80534	44	U 9000 / L 60135	72	U 70409 / L 30909
35	U 80534 / L 9000	45	U 60135 / L 9000	73	U 80534 / L 20900
36	U 41280 / L 30856	46	U 9000 / L 10927		
36A	U 41280 / L 41378	47	U 10927 / L 9000		

1954 MERCURY

01	India Black	9000
02	Atlantic Blue Poly	10993
03	Lakeland Blue	11364
04	Mohawk Maroon Poly	50326
05	Granby Gray	31086
08	Country Club Tan	20948
10	Bloomfield Green Poly	41328
11	Brentwood Brown Poly	20900

12	Glenoaks Green Poly	41484
15	Parklane Green	41364
17	Bittersweet	60135
18	Yosemite Yellow	80534
21	Siren Red	70409
22	Arctic White (Gray)	31121

TWO-TONES

28	U 41328 / L 41364	44	U 31121 / L 11364	66	U 60135 / L 31121
29	U 31121 / L 41364	51	U 20948 / L 20900	67	U 9000 / L 60135
30	U 31121 / L 41484	60	U 10993 / L 20948	68	U 9000 / L 80534
31	U 31121 / L 41328	61	U 10993 / L 31086	69	U 41484 / L 80534
36	U 41328 / L 20948	62	U 70409 / L 31086	70	U 41328 / L 80534
37	U 9000 / L 70409	63	U 31121 / L 9000	71	U 10993 / L 11364
40	U 31121 / L 10993	64	U 31121 / L 70409		
41	U 10993 / L 31121	65	U 31121 / L 60135		

1955 MERCURY

01	Tuxedo Black	9000
03	Biltmore Blue	11188
04	Gulfstream Blue Poly	11197
05	Kingston Gray	31198
07	Rockdale Gray Poly	31074
08	Forester Green Poly	41594
14	Springdale Green	41612
15	Canyon Cordovan Poly	60144
16	Lime	41643
17	Tropic Blue	11023
18	Yukon Yellow	80692
20	Arbor Green	41611
22	Alaska White	31121
24	Glen Lake Blue Poly	11146
26	Sea Isle Green Poly	41547
29	Carmen Red	70409
31	Persimmon	60208
203	Sunglaze	60209

TWO-TONES

32	U 41594 / L 41612	66	U 70409 / L 31121	82	U 11023 / L 31121
33	U 41612	67	U 11197	83	U 31121

#	Upper	Lower	#	Upper	Lower	#	Upper	Lower
		L 41594			L 11188			L 41611
35	U 31121	L 41547	68	U 11188	L 11197	84	U 41611	L 31121
36	U 41547	L 31121	69	U 31121	L 11197	85	U 31121	L 11146
38	U 31121	L 41594	70	U 11197	L 31121	86	U 11146	L 31121
39	U 41594	L 31121	71	U 31074	L 31198	87	U 9000	L 41643
40	U 31121	L 41612	72	U 31121	L 41643	88	U 31121	L 11188
41	U 9000	L 70409	77	U 9000	L 80692	94	U 31121	L 60208
62	U 41594	L 80692	78	U 31121	L 60144	95	U 31074	L 60208
63	U 31121	L 9000	79	U 60144	L 31121	201	U 41527	L 41612
64	U 9000	L 31112	80	U 31121	L 60209	202	U 41612	L 41547
65	U 31121	L 70409	81	U 31121	L 11023			

#	Upper	Lower	#	Upper	Lower	#	Upper	Lower
		L 70671			L 41643			L 41643
62	U 8050	L 31074	75	U 9000	L 8050	88	U 8050	L 41411
63	U 11396	L 11397	76	U 8050	L 9000	89	U 8050	L 80774
64	U 11397	L 11396	77	U 8050	L 41833	94	U 60208	L 31074
65	U 8050	L 11397	78	U 41833	L 8050	95	U 31074	L 60208
66	U 11397	L 8050	79	U 8050	L 41832	96A	U 9000	L 70671
67	U 8050	L 11396	80	U 8050	L 60209			

FLO-TONES

#	Code	#	Code	#	Code
207	11397	267	11396	281	60209
	11580		8050		8050
	11397		11396		60209
234	9000	268	41832	286	9000
	80774		41831		41411
	9000		41832		9000
235	80774	269	41831	287	9000
	9000		41832		41643
	80774		41831		9000
240	11396	270	41832	288	41411
	11406		41833		8050
	11396		41832		41411
241	11406	271	41833	288	41411
	11396		41832		8050
	11406		41833		41411
255A	70671	272	41643	294	60208
	8050		8050		31074
	70671		41643		60208
262	31074	275	9000	295	31074
	8050		8050		60208
	31074		9000		31074
263	11396	278	41833	296A	9000
	11397		8050		70671
	11396		41833		9000
264	11397	279	41832		
	11396		8050		
	11397		41832		
266	11397	280	8050		
	8050		60209		
	11397		8050		

1956 MERCURY

01	Tuxedo Black	9000
03	Delta Blue Poly	11406
04	Landerdale Blue	11397
05	Niagara Blue	11396
05A	Tyrolian Blue	11580
07	London Gray Poly	31074
08	Pinewood Green Poly	41831
14	Heath Green	41833
15	Glamour Tan	60209
17	Grove Green	41643
18	Verona Green	41832
18A	Spring Valley Green	42028
19	Saffron Yellow	80774
20	Cambridge Green	41411
21A	Carousel Red	70671
23	Classic White	8050
31	Persimmon	60208

TWO-TONES

#	Upper	Lower	#	Upper	Lower	#	Upper	Lower
34	U 9000	L 80774	68	U 41832	L 41831	81	U 60209	L 8050
40	U 11396	L 11406	69	U 41831	L 41832	82	U 8050	L 60208
41	U 11406	L 11396	70	U 41832	L 41833	83	U 60208	L 8050
55A	U 70671	L 8050	71	U 41833	L 41832	86	U 9000	L 41411
56A	U 8050		72	U 8050		87	U 9000	

1957 MERCURY

01	Tuxedo Black	9000
06	Classic White	8050
09	Tahitian Green (Blue)	11578
21	Regency Gray Poly	31412

22	Sherwood Green Poly	42038
23	Pacific Blue Poly	11585
24	Nantucket Blue	11562
25	Fiesta Red	70720
26	Brazilian Bronze Poly	21337
27	Pastel Peach	21336
28	Desert Tan	21318
29	Persimmon	60208
30	Rosewood	70723
31	Lexington Green	41890
32	Sunset Orchid	50443
05A	Tyrolian Blue	11580
18A	Spring Valley Green	42028
000	Sunglitter (Used on Pace Car)	80941

TWO-TONE COMBINATIONS

70	U 8050	145	U 9000	163	U 31412
	L 11578		L 11562		L 21318
71	U 11578	146	U 8050	164	U 8050
	L 8050		L 42038		L 21318
74	U 9000	147	U 42039	165	U 21318
	L 11578		L 8050		L 8050
75	U 9000	148	U 11578	166	U 80874
	L 80874		L 42038		L 31412
76	U 80874	149	U 42038	167	U 31412
	L 9000		L 11578		L 80874
77	U 8050	150	U 21336	168	U
	L 80874		L 70723		L 70720
78	U 80874	151	U 70723	169	U 70720
	L 8050		L 21336		L 8050
79	U 8050	152	U 21336	173	U 11578
	L 9000		L 9000		L 9000
80	U 9000	153	U 9000	181	U 9000
	L 8050		L 21336		L 70720
36	U 8050	154	U 21336	182	U 70720
	L 31412		L 21318		L 9000
37	U 31412	155	U 60208	183	U 70720
	L 8050		L 21336		L 9000
38	U 8050	156	U 21336	184	U 9000
	L 11585		L 21337		L 70723
39	U 11585	157	U 21337	185	U 8050
	L 8050		L 21336		L 70723
40	U 11562	158	U 8050	186	U 70723
	L 11585		L 21337		L 8050
41	U 11585	159	U 21337	187	U
	L 11562		L 8050		L 41890
42	U 8050	160	U 21318	188	U 41890
	L 11562		L 9000		L 8050
43	U 11562	161	U 9000	189	U 8050
	L 11578		L 21318		L 50443

44	U 11562	162	U 21318	190	U 50443
	L 9000		L 21318		L 8050
191	U 21336	194	U 60208	236	U 8050
	L 60208		L 8050		L 31412
192	U 60208	195	U 50443	237	U 31412
	L 21336		L 9000		L 8050
193	U 8050	196	U 9000		
	L 60208		L 50443		

FLO-TONE COMBINATIONS

238	8050	254	21336	270	8050
	11585		21318		11578
	8050		21336		8050
239	11585	255	21318	271	11578
	8050		21336		8050
	11585		21318		11578
240	11562	256	21336	273	11578
	11585		21337		9000
	11562		21336		11578
241	11585	257	21337	274	9000
	11562		21336		11578
	11585		21337		9000
242	8050	258	8050	275	9000
	11562		21337		80874
	8050		8050		9000
243	11562	259	21337	276	80874
	8050		8050		9000
	11562		21337		80874
244	11562	260	21318	277	8050
	9000		9000		80874
	11562		21318		8050
245	9000	261	9000	278	80874
	11562		21318		8050
	9000		9000		80874
246	8050	262	21318	279	8050
	42038		31412		9000
	8050		21318		8050
247	42038	263	31412	280	9000
	8050		21318		8050
	42038		31412		9000
248	11578	264	8050	281	9000
	42038		21318		70720
	11578		8050		9000
249	42038	265	21318	282	70720
	1157		88050		9000
	42038		21318		70720
250	21336	266	80874	283	70723
	70723		31412		9000
	21336		80874		70723
251	70723	267	31412	284	9000
	21336		80874		70723

	70723		31412		9000
252	21336	268	8050	285	8050
	9000		70720		70723
	21336		8050		8050
253	9000	269	70720	286	70723
	21336		8050		8050
	9000		70720		70723
287	8050	291	21336	295	50443
	41890		60208		9000
	8050		21336		50443
288	41890	292	60208	296	9000
	8050		21336		50443
	41890		60208		9000
289	8050	293	8050		
	50443		60208		
	8050		8050		
290	50443	294	60208		
	8050		8050		
	50443		60208		

1958 MERCURY

01	Tuxedo Black	9000
07	Marble White	8150
15	Parisian Green	42119
16	Emerald Poly	42169
17	Holly Green Poly	42117
30	Vineyard Blue	11699
31	Jamaican Blue Poly	11700
45	Flamingo Red	70799
55	Silver Sheen Poly	31608
56	Oxford Gray Poly	31581
66	Autumn Beige	21522
75	Mayfair Yellow	80944
87	Shadow Rose	70800
92	Golden Dust Poly	21521
97	Twilight Turquoise	11798
99	Burgundy Poly	50496

SINGLE TONE
101 = Tuxedo Black, 9000, with projectile area same as body color.
601 = Tuxedo Black, 9000, with projectile area in a contrasting color.

TWO-TONE
20107 = Tuxedo Black, 9000 and Marble White, 8150 with projectile area same as body color.
70107 = Tuxedo Black, 9000 and Marble White, 8150 with projectile area in a contrasting color.

TRI-TONE
30107 = Tuxedo Black, 9000 and Marble White, 8150 with projectile area same as body color.

80107 = Tuxedo Black, 9000 and Marble White, 8150 with projectile area in a contrasting color.

FLO-TONE
40107 = Tuxedo Black, 9000 and Marble White, 8150 with projectile area same as body color.
90107 = Tuxedo Black, 9000 and Marble White, 8150 with projectile area in a contrasting color.

1959 MERCURY

01	Tuxedo Black	9000
07	Marble White	8150
08	Glacier White	8103
16	Sagebrush Green Poly	42435
17	Sherwood Green Poly	42439
27	Satellite Blue	11848
28	Blue Ice Poly	11955
45	Canton Red	70794
55	Silver Sheen Poly	31608
56	Charcoal Poly	31797
66	Autumn Smoke	21673
75	Madeira Yellow	81059
83	Bermuda Sand	70926
84	Silver Beige Poly	70931
88	Twilight Turquoise	42434
92	Golden Beige Poly	21681
97	Neptune Turquoise Poly	11941

SINGLE TONE
01,07,17, ETC.

TWO-TONE STYLE #1
Projectile Area Same as Body Color
01-07
01 = Roof, Tuxedo Black 9000
07 = Balance of Car, Marble White 8150

TWO-TONE STYLE #2
7-01-07
7 = Projectile Area in Contrast to Surrounding Body Color.
01 = Roof and Projectile Area, Tuxedo Black 9000
07 = Upper and Lower, Marble White

TWO-TONE STYLE #3
7-01-07
7 = Projectile Area in Contrast to Surrounding Body Color
01 = Projectile Area, Tuxedo Black 9000
07 = Upper and Lower, Marble White 8150

1960 MERCURY AND COMET

A	Tuxedo Black	9000
B	Marine Blue Poly	12144
*C	Crystal Turquoise	42434

D	Aztec Turquoise Poly	11921
E	Cote D'azur Blue Poly	12236
F	Inlet Blue	12147
H	Javelin Bronze Poly	21857
J	Signal Red	71054
*K	Twilight Turquoise Poly	12238
M	Sultana White	8238
N	Polynesian Beige	21828
*Q	Royal Lilac Poly	50570
R	Sun Haze Yellow	81052
T	Valley Green Poly	42344
U	Mountain Rose Poly	21849
V	Summer Rose	71055
*W	Cameo Green	42634
X	Tucson Turquoise	12146
Z	Cloud Silver Poly	31991

*Comet only

In two-tone combinations the first letter indicates the lower color, the second letter upper.

1961 MERCURY AND COMET

A	Presidential Black	9000
*C	Turquoise Mist	42434
D	Blue Haze	12355
E	Saxon Green Poly	42820
*F	Sunburst Yellow	81249
*G	Tawny Beige	21982
H	Empress Blue Poly	12366
J	Signal Red	71054
K	Golden Bronze Poly	21981
*L	Gold Dust Poly	21886
M	Sultana White	8238
Q	Sheffield Gray Poly	32089
R	Columbia Blue Haze Poly	12361
S	Green Frost	42812
*V	Summer Rose	71055
W	Regency Turquoise Poly	12359

* Mercury only.

In two-tone combinations the first letter indicates the lowercolor, the second letter upper.

1962 MERCURY - MONTEREY - COMET - METEOR

A	Presidential Black	9000
B	Peacock	12676
D	Ocean Turquoise Poly	12489
E	Pacific Blue Poly	12494
F	Sea Blue	12488
H	Blue Satin Poly	121496
I	Castilian Gold Poly	32277
J	Carnival Red	71243
K	Light Aqua	42919

*L	Teaberry	71239
M	Sultana White	8238
P	Scotch Green Poly	42929
Q	Sheffield Gray Poly	32089
R	Jamaica Yellow	81324
T	Champagne	22110
*U	Velvet Turquoise Poly	12493
X	Black Cherry Poly	50593
Z	Desert Frost Poly	21958

* Monterey only.

In two-tone combinations the first letter indicates the lower color, the second letter upper.

1963 MERCURY - MONTEREY - COMET - METEOR

A	Presidential Black	9000
B	Peacock Turquoise	12676
D	Ocean Turquoise Poly	12489
E	Pacific Blue Poly	12494
F	Pink Lustre	71365
H	Blue Satin Poly	12496
I	Castilian Gold Poly	32277
J	Carnival Red	72143
M	Sultana White	8238
P	Scotch Green Poly	42929
R	Jamaica Yellow	81324
T	Champagne	22110
W	Pink Frost Poly	71322
X	Black Cherry Poly	50593
Y	Cascade Blue	12617
Z	Desert Frost Poly	21958

In two-tone combinations the first letter indicates the lower color, the second letter upper.

1964 MERCURY - PARKLANE - MONTCLAIR - MONTEREY - COMET

A	Onyx	9000***
B	Peacock	12851
D	Silver Turquoise	12853
F	Pacific Blue Poly	12832***
G	Palomino	22230
I	Aztec Gold	22395*
J	Carnival Red	71243***
K	Anniversary Silver Poly	32377
L	Bittersweet	60398*
M	Polar White	8378***
P	Pecan Frost Poly	22438
R	Yellow Mist	81444
S	Cypress Green Poly	42925***
T	Fawn	22249
V	Maize	81467
W	Pink Frost Poly	71322*

X	Burgundy Poly	50657
Y	Glacier Blue	22393***

* Mercury only.

** Comet only.

*** Accent color to match roof on two-tones

In two-tone combinations the first letter indicates lower color, the second letter upper.

1965 MERCURY - PARKLAND - MONTCLAIR - MONTEREY - COMET

A	Onyx	9000,9300
C	Ivy Gold Poly	22581
D	Silver Turquoise Poly	12853
F	Tiffany Blue	12854
H	Midnight Blue Poly	12547
I	Sandrift Poly	22436
J	Carnival Red	71243
K	Pearl Gray Poly	32377
M	Polar White	8378
O	Aquamarine	12852
P	Pecan Frost Poly	22438
R	Olive Mist Poly	43337
T	Fawn	22249
X	Burgundy Poly	50657
Y	Blue Ice Poly	12164
5	Ocean Turquoise Poly	12893
7	Yellow Mist	81444
8	Jamaican Yellow	81510

In two-tone combinations the first letter indicates lower color, the second letter upper.

1966 MERCURY - PARKLANE - MONTCLAIR - MONTEREY - COMET

A	Onyx	9000,9300
C	Coventry Gray Poly	32845
F	Tiffany Blue	12854
H	Sandstone	22528
K	Caspain Blue Poly	13076
M	Polar White	8378
P	Bronze Poly	22603
R	Olive Mist Poly	43408
T	Cardinal Red	71528
U	Turquoise Frost Poly	12745
V	Emberglo Poly	22610
X	Burgundy Poly	50669
Y	Blue Ice Poly	13045
Z	Sage Gold Poly	43433
2	Palisade Turquoise Poly	43454
4	Sheffield Silver Poly	32520
8	Jamaican Yellow	81510

In two-tone combinations the first letter or digit indicates lower color, the second upper color.

1967 MERCURY - PARKLANE - MONTCLAIR MONTEREY - COMET - COUGAR

A	Onyx	9000,9300
B	Turquoise	12876
D	Nordic Blue Poly	13357
E	Cumberland Beige Poly	22711
F	Tiffany Blue	12854
I	Lime Frost Poly	43576
K	Caspain Blue Poly	13076
M	Polar White	8378
O	Sea Foam Green	43529
Q	Glacier Blue Poly	12843
T	Cardinal Red	71528
V	Cinnamon Frost Poly	22749
W	Trafalgar Blue Poly	13073
X	Burgundy Poly	50669
Y	Inverness Green Poly	43567
Z	Sage Gold Poly	43433
4	Sheffield Silver Poly	32520
6	Fawn	22249
8	Jamaican Yellow	81510

In two-tone combinations the first letter or digit indicates lower color, the second letter or digit indicates the upper color.

1968 MERCURY

A	Onyx	9000,9300
B	Black Cherry	50746
*D	Nordic Blue Poly	13357
F	Madras Blue Poly	13329
I	Lime Frost Poly	43576
M	Polar White	8378
N	Diamond Blue	11683
O	Sea Foam Green	43529
P	Pewter Beige Poly	22744
Q	Glacier Blue Poly	13619
R	Augusta Green Poly	43644
T	Cardinal Red	71528
U	Caribbean Blue Poly	12745
W	Saxony	81584
X	Wellington Blue Poly	13356
Y	Grecian Gold Poly	22833
2	Tahitian Rose Poly	22737
3	Calypso Coral	60449
6	Fawn	22249

In two-tone combinations the first letter or digit indicates lower color, the second letter or digit indicates the upper color.

1969 MERCURY

B	Maroon	50746
C	Dark Ivy Green Poly	2037
*D	Pastel Gray	2038
E	Light Aqua	2039
F	Dark Aqua Poly	13329
H	Light Green	43575
I	Medium Lime Poly	2054
K	Dark Orchid Poly	2041
M	White	8378, 8734#2
*P	Medium Blue Poly	2042
Q	Medium Blue Poly	13619
S	Medium Gold Poly	2044
T	Red	71528
W	Yellow	81614
A	Black	900,9300
X	Dark Blue Poly	2056
Y	Burnt Orange Poly	2046
2	Light Ivy Yellow	2047
+*3	Competition Orange	60449
*4	Medium Green Poly	2048
*6	Bright Blue Poly	13357
8	Light Blue	2050
**9	Yellow	2052
***A	Black	9343
***B	Dark Blue Poly	13816
***D	Dark Red Poly	71753
***G	Dark Ivy Poly	44031
***K	Dark Aqua Poly	13817

* Cougar
** Cougar Spring Color
+ Also Montego Spring Color
*** Low Gloss - Marauder two-tone

1970 MERCURY

B	Dark Maroon	2150
C	Dark Green Poly	2146
D	Bright Yellow	2214
F	Dark Bright Aqua Poly	13329
*G	Medium Lime Poly	2152
J	Competition Blue	2230
K	Bright Gold Poly	2156
L	Light Gray Poly	2144
A	Black	9000,9300
M	White	8378
N	Pastel Blue	11683
P	Medium Ivy Green Poly	2147
Q	Medium Blue Poly	2138
S	Medium Gold Poly	2044
T	Red	71528

U	Competition Gold	2232
W	Yellow	2157
X	Dark Blue	2139
Y	Medium Bronze Poly	2142
Z	Competition Green	2231
1	Competition Orange	60449
2	Light Ivy Yellow	2047
5	Medium Brown Poly	2160
*6	Bright Blue Poly	13357
8	Light Gold	2043
*9	Pastel Yellow	81584
**A	Black	9343
**G	Dark Ivy	44031
**D	Red	71753

** Low Gloss Marauder two-tone
In two-tone combinations the first letter or digit indicates lower color; the second letter or digit indicates upper color For special colors furnish DSO or PTO number and serial number from warranty plate.

1971 MERCURY

A	History Onyx Black	9000,9300
B	Maroon Poly	2295
C	Dark Green Poly	2291
D	Competition Yellow	2214
E	Medium Yellow Gold	2299
F	Medium Blue Poly	2404
H	Light Green	2290
I	Bright Lime Green	2294
J	Competition Blue	2230
M	Knight White	8378
N	Pastel Blue	11683
P	Medium Green Poly	2289
Q	Medium Blue Poly	2372
S	Gray Gold Poly	2326
T	Red	71528
U	Competition Gold	2232
V	Light Pewter Poly	2287
W	Yellow	2157
Y	Dark Blue Poly	13076
Y	Dark Blue Poly	2405
Z	Competition Green Poly	2293
1	Competition Orange	60449
2	Bright Yellow	2414
3	Bright Red	2296
5	Medium Brown Poly	2371
6	Bright Blue Poly	13357
7	Maroon	50746
8	Light Gold	2043
9	Pastel Yellow	81584
49	Ivy Bronze Poly	2362
79	Ginger Bronze Poly	2363

125

1971 MONTEGO "SPOILER"

A	Black	9381
B	Dark Blue Poly	13816
G	Bright Yellow	44031

1972 MERCURY

1-A	Light Gray Poly	2409
2-B	Bright Red	2296
2-E	Red	71528
2-J	Maroon	50746
3-B	Light Blue	2403
3-D	Medium Blue Poly	2404
3-F	Competition Blue	2230
3-H	Dark Blue Poly	2405
3-J	Bright Blue Poly	13357
3-K	Blue Glamour Poly	2499
4-B	Bright Green Gold Poly	2406
4-C	Ivy Glamour Poly	2362
4-E	Bright Lime	2385
4-F	Medium Lime Poly	2412
4-P	Medium Green Poly	2289
4-Q	Dark Green Poly	2291
4-S	Light Green	2419
5-A	Light Pewter Poly	2287
5-H	Medium Brown Poly	2371
5-H	Ginger Poly	2482
5-J	Ginger Glamour Poly	2363
6-B	Light Goldenrod	2298
6-C	Medium Bright Yellow	2299
6-D	Yellow	2157
6-E	Medium Bright Yellow	2414
6-F	Gold Glamour Poly	2415
6-J	Gray Gold Poly	2326
9-A	White	8378
1-C	Black	9000,9300

In two-tone combinations the first two digits indicate lowercolor; the second two digits indicate upper color.

1973 MERCURY

1-G	Silver Poly	2593
2-B	Bright Red	2296
2-C	Red Poly	2428
2-L	Bright Red Poly	2574
3-B	Light Blue	2403
3-D	Medium Blue Poly	2404
3-K	Blue Glamour Poly	2499
3-M	Silver Glamour Poly	2501
3-N,P	Light Grabber Blue	2611
3-Q	Pastel Blue	2613
4-B	Green Gold Poly	2406
4-C	Ivy Glamour Poly	2362

4-N	Medium Aqua	2507
4-P	Medium Green Poly	2289
4-Q	Dark Green Poly	2291
4-S	Light Green	2419
5-A	Light Pewter Poly	2287
5-H	Ginger Poly	2482
5-J	Ginger Glamour Poly	2363
5-L	Tan	2285
5-M	Medium Copper Poly	2504
5-T	Saddle Bronze Poly	2575
6-B	Light Goldenrod	2298
6-C	Medium Goldenrod	2299
6-D	Yellow	2157
6-E	Medium Bright Yellow	2414
6-F	Gold Glamour Poly	2415
6-L	Medium Gold Poly	2497
9-A	White	8378
9-C	Special White	2512
1-C	Black	9000,9300

In two-tone combinations the first two digits indicate lower color; the second two digits indicate upper color.

1974 MERCURY

1-G	Silver Poly	2593
1-H	Medium State Blue Poly	2612
2-B	Bright Red	2296
2-E	Red	71528
3-B	Light Blue	2403
3-D	Medium Blue Poly	2404
3-E	Bright Blue Poly	2610
3-G	Dark Blue Poly	2472
3-M	Silver Blue Glamour Poly	2501
3-N	Light Grabber Blue	2611
3-Q	Pastel Blue	2613
4-A	Pastel Lime	2411
4-B	Green Gold Poly	2406
4-Q	Dark Green Poly	2291
4-T	Medium Ivy Bronze Poly	2666
4-V	Dark Yellow Green Poly	2614
4-W	Medium Lime Yellow	2615
5-H	Ginger Poly	2482
5-J	Ginger Glamour Poly	2363
5-M	Medium Chestnut Poly	2504
5-T	Saddle Bronze Poly	2575
5-U	Tan Poly	2618
5-W	Orange	2576
5-Y	Dark Copper Poly	2620
6-C	Medium Goldenrod	2299
6-F	Gold Glamour Poly	2415
6-M	Medium Dark Gold Poly	2621
6-N	Maize Yellow	2622

9-A	Wimbledon White	8378
9-C	White Decor	2512
9-D	White	2684
1-C	Black	9000,9300

In two-tone combinations the first two digits indicate lower color; the second two digits indicate upper color.

1975 MERCURY

1-G	Silver Poly	2593
1-H	Medium Slate Blue Poly	2612
2-B	Bright Red	2296
2-E	Red	71528
2-M	Dark Red	2609
2-Q	Maroon Poly	2716
3-G	Dark Blue Poly	2472
3-K	Bright Blue Glamour Poly	2499
3-M	Silver Blue Glamour Poly	2501
3-Q	Medium Pastel Blue	2613
4-T	Ivy Bronze Glamour Poly	2666
4-V	Dark Yellow Green Poly	2614
4-Z	Light Green Gold Poly	2720
5-J	Ginger Glamour Poly	2363
5-M	Medium Copper Poly	2504
5-Q	Dark Brown Poly	2616
5-T	Saddle Bronze Poly	2575
5-U	Tan Glamour Poly	2618
5-W	Orange	2576
5-Y	Dark Copper Poly	2620
6-D	Yellow	2157
6-E	Bright Yellow	2414
8-C	Medium Gold Poly	2813
9-D	White	2684
47	Light Green	2726
1-C	Black	9000,9300

In two-tone combinations the first two digits indicate lower color; the second two digits indicate upper color.

1976 MERCURY

1-G	Silver Poly	2593
1-H	Medium Slate Blue Poly	2612
2-B	Bright Red	2296
2-M	Dark Red	2609
2-R	Bright Red	2830
3-E	Bright Blue Poly	2610
3-G	Bright Dark Blue Poly	2472
3-M	Silver Blue Glamour Poly	2501
3-S	Light Blue	2834
4-T	Med. Ivy Bronze Glamour Poly	2666
4-V	Dark Yellow Green Poly	2614
5-M	Medium Chestnut Poly	2504

5-Q	Dark Brown Poly	2616
5-T	Saddle Bronze Poly	2575
5-U	Tan Glamour Poly	2618
6-E	Bright Yellow	2414
6-P	Cream	2790
6-T	Med. Drk. Gold Glamour Poly	2835
6-U	Tan	2836
6-V	Med. Gold Glamour Poly	2837
6-W	Light Gold	2838
8-C	Medium Gold Poly	2813
9-D	White	2684
46	Dark Jade Poly	2725
47	Light Green	2726
1-C	Black	9000,9300

In two-tone combinations the first two digits indicate lower color; the second two digits indicate upper color.

1977 MERCURY

1-G	Silver Poly	2593
*1-N	Dove Gray	2847
1-P	Med. Gray Poly (Two-Tone)	2967
2-M	Dark Red	2609
2-R	Bright Red	2830
2-U	Lipstick Red	2833
3-G	Brt. Drk. Blue Poly	2472
3-U	Light Blue	2907
3-V	Brt. Blue Poly	2908
4-V	Drk. Yellow Green Poly	2614
5-Q	Dark Brown Poly	2616
6-E	Bright Yellow	2414
6-P	Cream	2790
6-U	Light Tan	2836
6-V	Medium Gold Poly	2837
7-L	Light Jade Poly	2911
8-G	Vista Orange	2915
8-H	Tan	2916
8-J	Med. Tan Poly	2917
8-K	Bright Saddle Poly	2918
8-W	Chamois Poly	2923
8-Y	Champagne Poly	2924
9-D	White	2684
46	Dark Jade Poly	2725
47	Light Green	2726
1-C	Black	9000,9300

In two-tone combinations the first two colors indicate lower color; the second two digits indicate upper color.

1978 MERCURY

| 1-G | Silver Poly | 2593 |

1-N	Dove Gray	2847
1-P	Med. Gray Poly	2967
	(Two-Tone)	
2-M	Dark Red	2609
2-R	Bright Red	2830
2-U	Lipstick Red	2833
3-A	Dark Midnight Blue	3035
3-G	Brt. Drk. Blue Poly	2472
3-U	Light Blue	2907
3-V	Bright Blue Poly	2908
5-M	Medium Chestnut Poly	2504
5-Q	Dark Brown Poly	2616
6-E	Bright Yellow	2414
6-P	Cream	2790
6-W	Gold	2838
6-L	Medium Jade Poly	2911

7-W	Medium Jade	3044
8-J	Medium Tan Poly	2917
8-N	Dark Cord. Poly	2920
8-Y	Champagne Poly	2923
9-D	White	2684
21	Red	3060
34	Medium Blue	3048
46	Dark Jade Poly	2725
62	Antique Cream	3051
81	Russet Poly	3056
83	Light Chamois	3057
85	Tangerine	3058
1-C	Black	9000, 9300

In two-tone combinations the first two digits indicate lower color; the second two digits indicate upper color.

Chapter 12   OLDSMOBILE 1941-1978

PAINT CODE	COLOR	PPG CODE
1941 OLDSMOBILE		
20	Black	9200
21-A	Sea Plane Bronze Poly (Upper)	20045
	Casino Brown Poly (Lower)	
22	Ambassador Red Poly	20044
23	Miami Sand	20042
24	Capri Blue Poly	10083
24-A	Teal Blue Poly (Upper)	10084
	Capri Blue Poly (Lower)	10083
25	Oslo Blue Poly	10082
25-A	Pacific Blue Poly (Upper)	10085
	Oslo Blue Poly (Lower)	10082
26	Bengal Brown Poly	20043
27	Dusty Gray	30066
27-A	River Mist Gray (Upper)	30069
	Dusty Gray (Lower)	30066
28	Falcon Gray Poly	30065
28-A	Eddystone Gray Poly (Upper)	30067
	Falcon Gray Poly (Lower)	30065
29	Renfrew Green Poly	40084
29-A	Aspen Green Poly (Upper)	40085
	Renfrew Green Poly	40084
32	Garnet Red Poly	50014
1942 OLDSMOBILE		
40	Black	9200
41	Slate Green Poly	30065
42	Ambassador Red Poly	50019
43	New Ivory	80042
43-A	Warwick Tan Poly (Upper)	20054
	New Ivory (Lower)	80042
44	Darien Blue Poly	10097
44-A	Hurricane Blue (Upper)	10098
	Darien Blue Poly (Lower)	10097
45	Marine Blue Poly	10082
45-A	Swift Blue Poly (Upper)	10085
	Marine Blue Poly (Lower)	10082
46	Warwick Tan Poly	20054
46-A	Granada Brown (Upper)	20055
	Warwick Tan Poly (Lower)	20054
46-B	Warwick Tan Poly (Upper)	20054
	Granada Brown (Lower)	20055
47	Eagle Gray	30066
47-A	Condor Gray (Upper)	30069
	Eagle Gray (Lower)	30066

PAINT CODE	COLOR	PPG CODE
48	Condor Gray	30069
48-A	Black (Upper)	9200
	Condor Gray (Lower)	30069
49	Forest Green Poly	40084
49-A	Sea Foam Green Poly (Upper)	40085
	Forest Green Poly (Lower)	40084
49-B	Forest Green Poly (Upper)	40084
	Sea Foam Green Poly (Lower)	40085
49-C	Sea Foam Green Poly	40085
52	Garnet Red Poly	50014
1946 OLDSMOBILE		
60	Black	9200
61	Slate Green Poly	30065
62	Garnet Red Poly	50014
63-A	Pawnee Beige Poly (Upper)	20211
	New Ivory (Lower)	80042
63-C	New Ivory	80042
64	Tunis Blue Poly	10303
64-A	Nightshade Blue Poly (Upper)	10095
	Tunis Blue Poly (Lower)	10303
65	Nightshade Blue Poly	10095
65-B	Nightshade Blue Poly (Upper)	10095
	Pacific Blue Poly (Lower)	10085
66	Pawnee Beige Poly	20211
67-A	Channel Gray (Upper)	30312
	Eagle Gray (Lower)	30066
67-C	Eagle Gray	30066
68	Channel Gray	30312
68-A	Black (Upper)	9001
	Channel Gray (Lower)	30312
69	Forest Green Poly	40084
69-B	Forest Green Poly (Upper)	40084
	Sea Foam Green Poly (Lower)	40085
69-C	Sea Foam Green Poly	40085
1947 OLDSMOBILE		
70	Black	9200
71	Cambray Green Poly	40448
72	Ambassador Red Poly	50019
72-C	Chariot Red	70193
73-C	Havana Beige	20364
74	Caspian Blue Poly	10371
75	Nightshade Blue Poly	10095
76	Pawnee Beige Poly	20211
77-C	Saxon Gray Poly	30395
78	Chateau Gray Poly	30394
79	Ivy Green Poly	40098
79-C	Seafoam Green Poly	40085

TWO-TONES

71B	U 40448	75B	U 10095	
	L 30395		L 30395	
73A	U 20211	77A	U 30394	
	L 20364		L 30395	
74A	U 10095	79B	U 40098	
	L 10371		L 40085	

1948 OLDSMOBILE

20	Black	9200
21	Alpine Green Poly	40550
22-C	Chariot Red	70193
23	Tawnee Buff	20448
23-C	Nankeen Cream	80341
24	Denmark Blue Poly	10453
25	Caspian Blue Poly	10371
26	Praline Brown Poly	20449
27-C	Saxon Gray Poly	30395
28	Dawn Gray Poly	30490
29	Ivy Green Poly	40098
29-C	Seafoam Green Poly	40085
32	Garnet Red Poly	50125
34	Cayuga Blue Poly	10425
39-C	Norway Green	40551

TWO-TONES

21A	U 40551	28B	U 30490	
	L 40550		L 30395	
23A	U 20449	29A	U 40085	
	L 20448		L 40098	
24A	U 10371	29B	U 40098	
	L 10453		L 40085	
25A	U 10453	34A	U 10371	
	L 10371		L 10452	
26A	U 20448	34B	U 10452	
	L 20449		L 10371	
28A	U 30395	39A	U 40550	
	L 30490		L 40551	

1949 OLDSMOBILE

42	Garnet Maroon Poly	50125
42-C	Chariot Red	70193
43	Tawnee Buff	20562
44	Crest Blue	10555
45	Serge Blue Poly	10554
46	Praline Brown Poly	20449
47	Silver Gray	30614
48	Metal Gray Poly	30620
49	Ivy Green Poly	40098
49-C	Seafoam Green Poly	40085

TWO-TONES

43A	U 20449	47A	U 30620	
	L 20562		L 30614	
44A	U 10554	49B	U 40098	
	L 10555		L 40085	

1950 OLDSMOBILE

10	Black	9200
11	Adler Green Poly	40911
12	Garnet Maroon Poly	50125
12-C	Chariot Red	70193
13	Dune Beige	20681
13-C	Canto Cream	80472
14	Crest Blue	10555
15	Serge Blue Poly	10554
17	Marol Gray Poly	30741
18	Flint Gray Poly	30740
19	Ivy Green Poly	40098
29	Palm Green Poly	41048
41-C	Almond Green	40701
43-C	Nankeen Cream	80400
(Two-Tones only) Sand Beige		20751

TWO-TONES

11A	U 40098	17A	U 30740	
	L 40911		L 30741	
11H	U 9200	17H	U 9200	
	L 40911		L 30741	
13H	U 9200	29H	U 20751	
	L 80472		L 41048	
14A	U 10554			
	L 10555			

1951 OLDSMOBILE

50	Black	9200
51	Cascade Green Poly	41064
52	Empire Maroon Poly	50229
52-C	Chariot Red	70193
53	Sand Beige	20751
53-C	Canto Cream	80472
54	Otsego Blue Poly	10779
55	Serge Blue Poly	10554
55-C	Algiers Blue	10780
57	Dove Gray Poly	30870
58	Flint Gray Poly	30740
59	Palm Green Poly	41048
59-C	Shoal Green Poly	41065

TWO-TONES

51A	U 41048	54H	U 9200

	L 41064			L 10779
51H	U 9200		55H	U 9200
	L 41064			L 10780
52H	U 9200		57A	U 30740
	L 70193			L 30870
53H	U 9200		57H	U 9200
	L 80472			L 30870
54A	U 10554		59H	U 20751
	L 10779			L 41048

1952 OLDSMOBILE

10	Black	9200
20	Chariot Red	70193
21	Regent Maroon Poly	50285
21-A	Empire Maroon Poly	50229
30	Cascade Green Poly	41064
31	Shoal Green Poly	41065
31-W	Swan White	30934
32	Palm Green Poly	41048
33	Glade Green Poly	41143
40	Arctic Blue	10823
41	Serge Blue Poly	10554
50	Dove Gray Poly	30870
51	Pearl Gray Poly	30933
60	Canto Cream	80472
61	Sand Beige	20751
70	Aqua Marine Poly	10825
71	Royal Turquoise Poly	10828

TWO-TONES

20B	U 9200		40B	U 9200		60B	U 9200
	L 70193			L 10823			L 80472
30B	U 9200		40T	U 10554		71B	U 9200
	L 41064			L 10823			L 10828
30T	U 41048		40W	U 30934		71T	U 10825
	L 41064			L 10823			L 10828
31W	U 30934		50B	U 9200		71W	U 30934
	L 41056			L 30870			L 10828
32T	U 20751		50T	U 30933			
	L 41048			L 30870			
33T	U 20751		51W	U 30934			
	L 41143			L 30933			

1953 OLDSMOBILE

10	Black	9200
20	Agate Red	70419
21	Etna Maroon Poly	50334
30	Fern Green Poly	41321
31	Cove Green Poly	41322
32	Glade Green Poly	41143
40	Acacia Blue Poly	10948
41	Cadet Blue Poly	10949
42	Baltic Blue Poly	10947
50	Mist Gray Poly	31069
51	Pearl Gray Poly 1953	31068
60	Polar White	31070
61	Lotus Cream	80610
62	Monica Tan	20909
70	Royal Marine	10950
71	Regal Turquoise Poly	41320
(Two-Tones only)	Burma Brown Poly	20910

TWO-TONES

20-10	U 9200	30-31	U 41322	31-10	U 9200		
	L 70419		L 41321		L 41322		
20-60	U 31070	30-32	U 41143	31-31	U 41321		
	L 70419		L 41321		L 41322		
30-10	U 9200	30-60	U 31070	31-32	U 41143		
	L 41321		L 41321		L 41322		
31-60	U 31070	42-60	U 31070	60-51	U 31068		
	L 41322		L 10947		L 31070		
32-30	U 41321	50-10	U 9200	60-63	U 20910		
	L 41143		L 31069		L 31070		
32-31	U 41322	50-32	U 41143	60-70	U 10950		
	L 41143		L 31069		L 31070		
32-60	U 31070	50-42	U 10947	60-71	U 41320		
	L 41143		L 31069		L 31070		
40-10	U 9200	50-51	U 31068	61-63	U 20910		
	L 10948		L 31068		L 86010		
40-41	U 10949	51-10	U 9200	62-10	U 9200		
	L 10948		L 31068		L 20909		
40-42	U 10947	51-50	U 31069	62-63	U 20910		
	L 10948		L 31068		L 20909		
40-60	U 31070	51-60	U 31070	70-50	U 31069		
	L 10948		L 31068		L 10950		
41-10	U 9200	60-30	U 41321	70-60	U 31070		
	L 10949		L 31070		L 10950		
41-40	U 10948	60-31	U 41322	71-10	U 9200		
	L 10949		L 31070		L 41320		
41-42	U 10947	60-32	U 41143	71-60	U 31070		
	L 10949		L 31070		L 41320		
41-60	U 31070	60-40	U 10948	71-70	U 10950		
	L 10949		L 31070		L 41320		
42-40	U 10948	60-41	U 10949				
	L 10947		L 31070				
42-41	U 10949	60-42	U 10947				
	L 10947		L 31070				

1954 OLDSMOBILE

10	Black	9200
21	Etna Maroon Poly	50334
22	Flare Red	70485

30	Glacier Green	41562
31	Willow Green Poly	41566
32	Glade Green Poly	50334
33	Sarasota White (Green)	41673
40	Capri Blue	11148
41	Cadet Blue Poly	10949
42	Baltic Blue Poly	10947
50	Mist Gray Poly	31069
51	Juneau Gray Poly	31185
60	Polar White	31070
61	Maize Cream	80670
62	Desert Tan	21060
63	Copper Metallic Poly	21061
70	Royal Marine	10950
72	Turquoise Poly	11132

TWO-TONES

21-62	U 21060 L 50334	40-42	U 10947 L 11148	60-41	U 10949 L 31070
22-10	U 9200 L 70485	40-60	U 31070 L 11148	60-42	U 10947 L 31070
22-50	U 31069 L 70485	41-10	U 9200 L 10949	60-51	U 31185 L 31070
22-60	U 21070 L 70485	41-60	U 31070 L 10949	60-63	U 21061 L 31070
22-62	U 21060 L 70485	42-40	U 11148 L 10947	60-70	U 10950 L 31070
30-10	U 9200 L 41562	42-60	U 31070 L 10947	60-72	U 11132 L 31070
30-32	U 41143 L 41562	50-10	U 9200 L 31069	61-10	U 9200 L 80670
30-60	U 31070 L 41562	50-22	U 70485 L 31069	61-31	U 41566 L 80670
31-10	U 9200 L 41566	50-32	U 41143 L 31069	61-32	U 41143 L 80670
31-60	U 31070 L 41566	50-42	U 10947 L 31069	61-60	U 31070 L 80670
32-30	U 41562 L 41143	50-51	U 31185 L 31069	62-10	U 9200 L 21060
32-60	U 31070 L 41566	51-10	U 9200 L 31185	62-21	U 50334 L 21060
32-61	U 80670 L 41143	51-50	U 31069 L 31185	62-63	U 21061 L 21060
33-10	U 9200 L 41673	51-60	U 31070 L 31185	63-60	U 31070 L 21061
33-31	U 41566 L 41673	60-30	U 41562 L 31070	63-62	U 21060 L 21061
33-32	U 41143 L 41673	60-22	U 70485 L 31070	70-50	U 31069 L 10950
33-51	U 31185 L 41573	60-31	U 41566 L 31070	70-60	U 31070 L 10950
33-63	U 21061 L 41673	60-32	U 41143 L 31070	72-10	U 9200 L 11132
40-10	U 9200 L 11148	60-40	U 11148 L 31070	72-60	U 31070 L 11132

1955 OLDSMOBILE

10	Black	9200
20	Burlingame Red	70559
21	Regal Maroon	50403
30	Mint Green	41731
31	Glen Green Poly	41732
32	Grove Green Poly	41733
40	Twilight Blue	11304
41	Panama Blue Poly	11320
42	Bimini Blue Poly	11319
43	Frost Blue	11322
50	Mist Gray Poly	31069
51	Juneau Gray Poly	31185
60	Turquoise Poly	31070
61	Caspian Cream	80671
62	Coral	60191
63	Bronze Metallic Poly	21148
64	Chartreuse	41734
65	Shell Beige	21149
70	Turquoise Poly	11132

TWO-TONES

10-50	U 31069 L 9200	40-43	U 11322 L 11304	60-63	U 21148 L 31070
10-60	U 31070 L 9200	41-40	U 11304 L 11320	60-70	U 11132 L 31070
10-70	U 11132 L 9200	41-60	U 31070 L 11320	61-10	U 9200 L 80671
10-64	U 41734 L 9200	41-43	U 11322 L 11320	61-32	U 41733 L 80671
20-10	U 9200 L 70559	42-40	U 11304 L 11319	61-63	U 21148 L 80671
20-50	U 31069 L 70559	42-60	U 31070 L 11319	62-10	U 9200 L 60191
20-60	U 31070 L 70559	42-43	U 11322 L 11319	62-50	U 31069 L 60191
20-65	U 21149 L 70559	50-10	U 9200 L 31069	62-60	U 31070 L 60191
21-10	U 9200 L 50403	50-20	U 70559 L 31069	62-65	U 21149 L 60191
21-50	U 31069 L 50403	50-21	U 50403 L 31069	43-10	U 9200 L 11322
21-60	U 31070 L 50403	50-42	U 11319 L 31069	43-40	U 11304 L 11322
21-65	U 21149 L 50403	50-51	U 31185 L 31069	43-41	U 11320 L 11322

30-10 U 9200 / L 41731	51-50 U 31069 / L 31185	43-42 U 11319 / L 11322
30-31 U 41732 / L 41731	51-60 U 31070 / L 31185	70-10 U 9200 / 11132
30-32 U 41733 / L 41731	60-10 U 9200 / L 31070	70-60 U 31070 / L 9200
30-63 U 21148 / L 41731	60-20 U 70559 / L 31070	64-10 U 9200 / L 41734
31-10 U 9200 / L 41732	60-21 U 50403 / L 31070	64-51 U 31185 / L 41734
31-30 U 41731 / L 41732	60-31 U 41732 / L 31070	64-63 U 21148 / L 41734
31-60 U 31070 / L 41732	60-32 U 41733 / L 31070	65-20 U 70559 / L 21149
32-30 U 41731 / L 41733	60-40 U 11304 / L 31070	65-21 U 50403 / L 21149
32-60 U 31070 / L 41733	60-41 U 11320 / L 31070	65-62 U 60191 / L 21149
40-10 U 9200 / L 11304	60-42 U 11319 / L 31070	65-63 U 21148 / L 21149
40-41 U 11320 / L 11304	60-51 U 31185 / L 31070	63-60 U 41731 / L 21148
40-42 U 11319 / L 11304	60-61 U 80671 / L 31070	63-60 U 31070 / L 21148
40-60 U 31070 / L 11304	60-62 U 60191 / L 31070	63-65 U 21149 / L 21148

Special Two-Tones can be identified by the letter "S", followed by the paint code number. For example -

S-63-61 Front End & Upper Body	80671
Rear End & Lower Body	21148

1956 OLDSMOBILE

10	Black	9200
20	Festival Red	70669
30	Ice Green	41929
31	Canyon Green Poly	41927
32	Tropical Green Poly	41928
40	Cirrus Blue	11485
41	Artesian Blue Poly	11483
42	Nordic Blue Poly	11484
50	Sterling Gray Poly	31366
51	Juneau Gray Poly	31185
52	Charcoal Gray Poly	31367
60	Alcan White	8123
61	Citron Cream	80820
62	Terra Cotta	60229
63	Shantung Beige	21247
64	Citation Bronze Poly	21248
65	Lime	41930
66	Island Coral	60232
70	Turquoise Poly	11482

90	Antique White	8125
91	Gold Mist Poly	21366
92	Rose Mist Poly	70678

TWO-TONES

10-20 U 70669 / L 9200	20-63 U 21247 / L 70669	32-60 U 8123 / L 41928
10-40 U 11485 / L 9200	30-10 U 9200 / L 41929	32-61 U 80820 / L 41928
10-41 U 11483 / L 9200	30-31 U 41927 / L 41929	32-63 U 21247 / L 41928
10-50 U 31366 / L 9200	30-32 U 41928 / L 41929	40-10 U 9200 / L 11485
10-51 U 31185 / L 9200	30-52 U 31185 / L 41929	40-41 U 11483 / L 11485
10-60 U 8123 / L 9200	30-60 U 8123 / L 41929	40-42 U 11484 / L 11485
10-61 U 80820 / L 9200	31-10 U 9200 / L 41927	40-51 U 31185 / L 11485
10-62 U 60229 / L 9200	31-30 U 41929 / L 41927	40-52 U 31367 / L 11485
10-63 U 21247 / L 9200	31-32 U 41928 / L 41927	40-60 U 8123 / L 11485
10-65 U 41930 / L 9200	31-60 U 8123 / L 41927	41-40 U 11485 / L 11483
10-66 U 60232 / L 9200	31-61 U 80820 / L 41927	41-42 U 11484 / L 11483
10-70 U 11482 / L 9200	31-63 U 21247 / L 41927	41-50 U 31366 / L 11483
20-10 U 9200 / L 70669	32-30 U 41929 / L 41928	41-60 U 8123 / L 11483
20-50 U 31366 / L 70669	32-31 U 41927 / L 41928	42-40 U 11485 / L 11484
20-52 U 31367 / L 70669	32-50 U 31366 / L 41928	42-41 U 11483 / L 11484
20-60 U 8123 / L 70669	32-51 U 31185 / L 41928	42-50 U 31366 / L 11484
42-60 U 8123 / L 11484	52-65 U 41930 / L 31367	62-63 U 21247 / L 60229
50-10 U 9200 / L 31366	52-66 U 60232 / L 31367	63-10 U 9200 / L 21247
50-20 U 70669 / L 31366	52-70 U 11482 / L 31367	63-20 U 70669 / L 21247
50-31 U 41927 / L 31366	60-10 U 9200 / L 8123	63-31 U 41927 / L 21247
50-32 U 41928 / L 31366	60-20 U 70669 / L 8123	63-32 U 41928 / L 21247
50-41 U 11483 / L 31366	60-30 U 41929 / L 8123	63-62 U 60229 / L 21247
50-42 U 11484 / L 31366	60-31 U 41927 / L 8123	63-64 U 21248 / L 21247

Code	Upper	Lower	Code	Upper	Lower	Code	Upper	Lower
50-51	U 31185	L 31366	60-32	U 41928	L 8123	64-60	U 8123	L 21248
50-52	U 31367	L 31366	60-40	U 11485	L 8123	64-61	U 80820	L 21248
50-60	U 8123	L 31366	60-41	U 11483	L 8123	64-63	U 21247	L 21248
50-62	U 60229	L 31366	60-42	U 11484	L 8123	65-10	U 9200	L 41930
50-65	U 41930	L 31366	60-50	U 31366	L 8123	65-32	U 41928	L 41930
51-10	U 9200	L 31185	60-51	U 31185	L 8123	65-50	U 31366	L 41930
51-42	U 11484	L 31185	60-52	U 31367	L 8123	65-51	U 31185	L 41930
51-50	U 31366	L 31185	60-62	U 60229	L 8123	65-52	U 31367	L 41930
51-52	U 31367	L 31185	60-63	U 21247	L 8123	65-60	U 8123	L 41930
51-60	U 8123	L 31185	60-64	U 21248	L 8123	66-10	U 9200	L 60232
51-61	U 80820	L 31185	60-65	U 41930	L 8123	66-51	U 31185	L 60232
51-65	U 41930	L 31185	60-66	U 60232	L 8123	66-52	U 31367	L 60232
51-66	U 60232	L 31185	60-70	U 11482	L 8123	66-60	U 8123	L 60232
52-20	U 70669	L 31367	61-10	U 9200	L 80820	66-63	U 21247	L 60232
52-40	U 11485	L 31367	61-32	U 41928	L 80820	70-10	U 9200	L 11482
52-41	U 11483	L 31367	61-51	U 31185	L 80820	70-52	U 31367	L 11482
52-50	U 31366	L 31367	61-52	U 31367	L 80820	70-60	U 8123	L 11482
52-51	U 31185	L 31367	61-60	U 8123	L 80820	90-91	U 21366	L 8125
52-60	U 8123	L 31367	61-64	U 21248	L 80820	91-90	U 8125	L 21366
52-61	U 80820	L 31367	62-10	U 9200	L 60229	92-90	U 8125	L 70678
52-62	U 60229	L 31367	62-52	U 31367	L 60229			
52-63	U 21247	L 31367	62-60	U 8123	L 60229			

	Color	Code
1	Allegheny Green	42083
0	Banff Blue	11628
1	Artesian Blue Poly	11483
0	Grenada Gray	31456
1	Juneau Gray Poly	31185
2	Charcoal Poly	31457
0	Alcan White	8123
1	Coronado Yellow	80910
2	Sunset Glow	70749
3	Shantung Beige	21247
4	Cutlass Bronze Poly	21368
5	Desert Glow	70748
6	Royal Glow	70766
0	Victoria White	8143
1	Gold Mist Poly	21366
2	Rose Mist Poly	70678
3	Sapphire Mist Poly	11629
4	Jade Mist Poly	42084
5	Platinum Mist Poly	31458

Color combinations can be identified by the arrangement of the paint code numbers. For example -

SINGLE TONE

10,20,Etc.

TWO-TONES

10-60

"10" = Body Color, Black, 9000

"60" = Accent Color, Alcan White, 8123

THREE-TONES

60-20-10

"60" = Lower, Alcan White, 8123

"20" = Upper, Festival Red, 70669

"10" = Accent Color, Black, 9000

1957 OLDSMOBILE

	Color	Code
0	Black	9000
0	Festival Red	70669
1	Accent Red	70765
2	Accent Vermillion	60257
0	Ice Green	41929

1958 OLDSMOBILE

	Color	Code
	Onyx Black	9000
	Festival Red	70851
	Surf Green	42238
	Allegheny Green Poly	42083
	Banff Blue	11628
	Marlin Blue Poly	11763
	Pearl Gray	31637
	Sterling Gray Poly	31636
	Charcoal Poly	31638
60	Alaskan White	8123
63	Sandstone Poly	21531
64	Autumn Haze Poly	60269
65	Desert Glow	70748
66	Canyon Glow Poly	70833
67	Heather	50503
68	Mountain Haze Poly	50504
90	Victorian White	8143

91	Champagne Mist Poly	21530
92	Rose Mist Poly	70678
93	Turquoise Mist Poly	42237
94	Jade Mist Poly	42084
95	Tropical Mist Poly	11762

Color combinations can be identified by the arrangement of paint code numbers. For example -

SINGLE TONE
10,20,Etc.

TWO-TONES
10-20-10

"10" = Lower Color, Onyx Black, 9000
"20" = Upper Color, Festival Red, 70851
"10" = Wheel Color, Onyx Black, 9000

1959 OLDSMOBILE

A	Ebony Black	9300
B	Silver Mist Poly	31827
C	Polaris White	8160
D	Willow Mist Poly	42479
E	Emerald Mist Poly	42480
F	Crystal Green	42499
H	Sapphire Mist Poly	12002
J	Frost Blue	12003
K	Aqua Mist Poly	12001
L	Cardinal Red	70961
M	Russet Poly	70959
N	Burgundy Mist Poly	50536
P	Golden Mist Poly	21722
R	Bronze Mist Poly	21723
S	Indigo Poly	12025

NOTE: On two-tones and insert combinations the first letter is the lower color, the second center area or accent color, the third is the upper color and the fourth letter is the wheel color.

1960 OLDSMOBILE

A	Ebony Black	9300
B	Charcoal Mist Poly	31905
C	Provincial White	8259
D	Platinum Mist Poly	31928
F	Gulf Blue Poly	12174
H	Dresden Blue Poly	12234
J	Palmetto Mist Poly	42650
K	Fern Mist Poly	42693
L	Garnet Mist Poly	50568
M	Citron	81202
N	Cordovan Poly	21874
P	Golden Mist Poly	21722
R	Shell Beige	21873
S	Copper Mist Poly	21841

| T | Turquoise Poly | 12228 |

Two-tone combinations can be identified by the paint code letters, for example -
A = Ebony Black
B = Charcoal Mist Poly

1961 OLDSMOBILE & F-85

A	Ebony Black	9300
B	Twilight Mist Poly	50597
C	Provincial White	8259
D	Platinum Mist Poly	31928
F	Azure Mist Poly	12398
H	Glacier Blue	12399
J	Tropic Mist Poly	42837
K	Alpine Green	42838
L	Garnet Mist Poly	50568
N	Cordovan Mist Poly	21874
P	Turquoise Mist Poly	12396
Q	Aqua	12401
R	Sandalwood	21733
S	Autumn Mist Poly	71211
T	Fawn Mist Poly	22005

Two-tone combinations can be identified by the paint code letters for example -
A = Ebony Black
B = Twilight Mist Poly

1962 OLDSMOBILE

A	Ebony Black	9300
B	Heather Mist Poly	12551
C	Provincial White	8259
D	Sheffield Mist Poly	32173
F	Wedgewood Mist Poly	12546
H	Cirrus Blue	12549
J	Willow Mist Poly	42975
K	Surf Green	42974
L	Garnet Mist Poly	50568
M	Cameo Cream	81271
N	Royal Mist Poly	50615
P	Pacific Mist Poly	12525
R	Sand Beige	22137
T	Sahara Mist Poly	22121
V	Chariot Red	70961
X	Sunset Mist Poly	71269

Two-tone combinations can be identified by the paint code letters for example -
A = Ebony Black
B = Heather Mist Poly

1963 OLDSMOBILE

| A | Ebony Black | 9300 |

C	Provincial White	8259
D	Sheffield Mist Poly	32173
F	Wedgewood Mist Poly	12546
H	Cirrus Blue	12713
J	Willow Mist Poly	42975
K	Barktone Mist Poly	22294
L	Regal Mist Poly	50633
P	Pacific Mist Poly	12525
R	Sand Beige	22137
T	Sahara Mist Poly	22121
V	Chariot Red	70961
X	Sunset Mist Poly	71269

Two-tone combinations can be identified by the paint code letters for example -
A = Ebony Black, 9300
B = Heather Mist Poly, 12551, etc.

1963 OLDSMOBILE

A	Ebony Black	9300
C	Provincial White	8259
D	Sheffield Mist Poly	32173
F	Wedgewood Mist Poly	12546
H	Cirrus Blue	12713
J	Willow Mist Poly	42975
K	Barktone Mist Poly	22294
L	Regal Mist Poly	50633
P	Pacific Mist Poly	12525
R	Sand Beige	22137
S	Saddle Mist Poly	22269
T	Sahara Mist Poly	22268
V	Holiday Red	71336
W	Midnight Mist Poly	12696
X	Antique Rose Poly	71337

Paint combinations can be identified as follows:
First letter indicates lower body color.
Second letter indicates upper body color.
Third letter indicates wheel or deluxe disc color.

1964 OLDSMOBILE

A	Ebony Black	9300
C	Provincial White	8259
D	Sheffield Mist Poly	32173
E	Jade Mist Poly	43263
F	Wedgewood Mist Poly	12546
H	Bermuda Blue	12847
J	Fern Mist Poly	43264
K	Tahitian Yellow	81450
L	Regal Mist Poly	50633
L	Regal Mist Poly #2	50684
P	Pacific Mist Poly	12525
Q	Aqua Mist Poly	12848

R	Cashmere Beige	22391
S	Saddle Mist Poly	22269
V	Holiday Red	71336
W	Midnight Mist Poly	12696

In two-tone combinations the first letter indicates lower color, the second letter upper color.

1965 OLDSMOBILE

A	Ebony Black	9300
B	Nocturne Mist Poly	32448
C	Provincial White	8259
D	Lucerne Mist Poly	13042
E	Royal Mist Poly	13002
H	Laurel Mist Poly	43391
J	Forest Mist Poly	43390
K	Ocean Mist Poly	43364
L	Turquoise Mist Poly	13003
N	Burgundy Mist Poly	50700
R	Target Red	71472
T	Mojave Mist Poly	22564
V	Almond Beige	22270
W	Sterling Mist Poly	32461
Y	Saffron Yellow	81500

In two-tone combinations the first letter indicates lower color, the second letter upper color.

1966 OLDSMOBILE & F-85

A	Ebony Black	9300
B	Nocturne Mist Poly	32448
C	Provincial White	8259
D	Lucerne Mist Poly	13042
E	Royal Mist Poly	13002
G	Trumpet Gold Poly	22661
H	Laurel Mist Poly	43391
J	Forest Mist Poly	43390
K	Ocean Mist Poly	43364
L	Tropic Turquoise Poly	43496
M	Autumn Bronze Poly	71525
N	Burgundy Mist Poly	50700
R	Target Red	71472
S	Champagne Mist Poly	22662
T	Sierra Mist Poly	22660
U	Dubonnet	50722
V	Almond Beige	22270
W	Silver Mist Poly	32525
X	Porcelain White	8631
Z	Frost Green Poly	43525

In two-tone combinations the first letter indicates lower color, the second upper color.

1967 OLDSMOBILE

A	Ebony Black	9300
B	Turquoise Frost Poly	43664
C	Provincial White	8259
D	Crystal Blue Poly	13349
E	Midnight Blue Poly	13346
F	Bimini Blue Poly	13364
G	Gold Poly	22818
H	Aspen Green Poly	43651
J	Emerald Green Poly	43653
K	Aquamarine Poly	43661
L	Tahoe Turquoise Poly	43659
N	Burgundy Mist Poly	50700
P	Pewter Poly	32603
R	Spanish Red	71583
S	Champagne Poly	22813
T	Cameo Ivory	81578
U	Dubonnet	50722
V	Antique Pewter Poly	32604
W	Sauterne Poly	22821
X	Garnet Red Poly	71585
Y	Saffron	81500
Z	Florentine Gold Poly	43665

1968 OLDSMOBILE

A	Ebony Black	9300
B	Twilight Teal Poly	13515
C	Provincial White	8259
D	Sapphire Blue Poly	13512
E	Nocturne Blue Poly	13513
F	Teal Frost Poly	13514
G	Willow Gold Poly	22942
K	Ocean Turquoise Poly	13517
L	Teal Blue Poly	13516
M	Cinnamon Bronze Poly	22967
N	Burgundy Poly	50775
P	Silver Green Poly	43774
R	Scarlet	71634
S	Jade Gold Poly	43794
T	Ivory	81617
V	Juneau Gray Poly	43773
W	Silver Beige Poly	22962
X	Buckskin	22983
Y	Saffron	81500
Z	Peruvian Silver Poly	8596

1969 OLDSMOBILE

10	Ebony Black	9300
40	Saffron	81500
50	Cameo White	2058
51	Trophy Blue Poly	2075
52	Crimson	2076
53	Nassau Blue Poly	2077
55	Tahitian Turquoise Poly	2078
46-0	Toronado Jade Poly	2172
57	Glade Green Poly	2079
59	Meadow Green Poly	2080
61	Sable Poly	2081
63	Palomino Gold Poly	22813
65	Topaz Poly	2082
67	Burgundy Mist (Poly)	50700
69	Platinum Poly	2059
75	Aztec Gold Poly	2085
77	Autumn Gold Poly	2086
80	Powder Blue	2096
81	Flamingo Silver Poly	2087
82	Covert Beige	2088
83	Deauville Gray Poly	32526
85	Chestnut Bronze Poly	22663

TORONADO OPTION COLORS

1-01	Amethyst Poly	50828
U-02	Caribbean Turquoise Poly	13839
X-03	Nugget Gold Poly	23211

1970 OLDSMOBILE

10	Porcelain White	8631
19	Ebony Black	9300
14	Platinum Poly	2059
16	Oxford Gray Poly	2161
20	Azure Blue	2162
25	Astro Blue Poly	2165
26	Viking Blue Poly	2213
28	Twilight Blue Poly	2166
34	Reef Turquoise Poly	2168
38	Aegean Aqua Poly	2234
45	Aspen Green Poly	2171
46	Ming Jade Poly	2172
48	Sherwood Green Poly	2173
50	Bamboo	2175
51	Sebring Yellow	2094
53	Nugget Gold Poly	23211
55	Galleon Gold Poly	2178
58	Burnished Gold Poly	2179
61	Sandalwood	2181
63	Copper Poly	2183
68	Cinnamon Bronze Poly	2233
73	Rally Red	2235
74	Grenadier Red	71642
75	Matador Red	2189
76	Regency Rose Poly	2190
78	Burgundy Mist Poly	50700

1971 OLDSMOBILE

11	Cameo White	2058
13	Sterling Silver Poly	2327
16	Oxford Gray Poly	2161
19	Ebony Black	9300
24	Nordic Blue Poly	2328
26	Viking Blue Poly	2213
29	Monarch Blue Poly	2330
39	Capri Aqua Poly	2331
41	Silver Mint Poly	2332
42	Palm Green Poly	2333
43	Lime Green Poly	2334
49	Antique Jade Poly	2337
50	Bamboo	2175
53	Saturn Gold Poly	2339
55	Galleon Gold Poly	2178
61	Sandalwood	2181
62	Bittersweet Poly	2340
65	Kashmir Copper Poly	2343
67	Sienna Poly	23215
68	Sable Brown Poly	2344
70	Doeskin Poly	2346
73	Autumn Glow Poly	2347
74	Venetian Red Poly	2348
75	Matador Red	2189
78	Antique Briar Poly	2350

1972 OLDSMOBILE

11	Cameo White	2058
19	Ebony Black	9300
14	Silver Pewter Poly	2429
18	Antique Pewter Poly	2430
24	Nordic Blue Poly	2328
26	Viking Blue Poly	2213
28	Royal Blue Poly	2166
36	Radiant Green Poly	2433
43	Pinehurst Green Poly	2435
48	Sequoia Green Poly	2439
50	Covert Beige	2441
53	Saturn Gold Poly	2339
54	Sovereign Gold Poly	2442
56	Sunfire Yellow	2444
57	Baroque Gold Poly	2445
62	Saddle Tan	2447
63	Saddle Bronze Poly	2448
65	Flame Orange Poly	2450
69	Nutmeg Poly	2452
75	Matador Red	2189
81	Bamboo	2175
A	White	8856
B	Black	9348

K	Black	9396
F	Medium Tan	23661
T	Light Covert	44544
G	Medium Green	44545

1973 OLDSMOBILE

11	Cameo White	2058
24	Wedgewood Blue Poly	2523
26	Zodiac Blue Poly	2524
29	Eclipse Blue Poly	2526
42	Emerald Green Poly	2528
44	Crystal Green Poly	2529
46	Moss Green Poly	2530
48	Brewster Green	2531
51	Omega Yellow	2533
56	Chamois Gold	2537
60	Mayan Gold Poly	2538
64	Silver Taupe Poly	2541
66	Tanbark Poly	2542
68	Chestnut Poly	2543
74	Cranberry Red Poly	2545
75	Omega Red	2546
81	Honey Beige	2549
85	Tiffany Gold Poly	2483
97	Omega Orange Poly	2555
19	Ebony Black	9300

In two-tone combinations the first two digits indicate lower color, the next two digits the upper color.

1974 OLDSMOBILE

11	Cameo White	2058
19	Ebony Black	9300
24	Wedgewood Blue Poly	2523
26	Zodiac Blue Poly	2524
29	Eclipse Blue Poly	2526
36	Reef Turquoise Poly	2640
40	Omega Lime	2641
44	Sage Green	2642
46	Cypress Green Poly	2643
49	Balsam Green Poly	2645
50	Colonial Cream	2646
51	Omega Maize	2677
53	Omega Gold Poly	2649
55	Colonial Gold	2650
59	Citation Bronze Poly	2367
64	Silver Taupe Poly	2541
66	Cinnamon Poly	2653
69	Clove Brown Poly	2656
74	Cranberry Poly	2658
75	Omega Red	2546

1975 OLDSMOBILE

11	Cameo White	2058
19	Ebony Black	9300
13	Inca Silver Poly	2518
15	Dove Gray	2742
16	Shadow Gray Poly	2743
21	Glacier Blue Poly	2431
24	Horizon Blue	2745
26	Spectre Blue Poly	2746
29	Midnight Blue Poly	2748
44	Sage Green	2642
49	Forest Green Poly	2752
50	Colonial Cream	2646
51	Sebring Yellow	2677
55	Sandstone	2755
58	Omega Bronze Poly	2757
59	Sable Brown Poly	2758
63	Canyon-Copper Poly	2759
64	Permission Poly	2760
72	Crimson Red	2544
74	Cranberry Poly	2658
75	Omega Red	2546
79	Burgundy Poly	2659
80	Sunfire Orange Poly	2548

In two-tone combinations the first two digits indicate lower color, the next two digits the upper color.

1976 OLDSMOBILE

11	Cameo White	2058
19	Ebony Black	9300
13	Inca Silver Poly	2518
16	Medium Gray Poly	2862
28	Light Blue Poly	2772
35	Dark Blue Poly	2863
36	Red Poly	2811
37	Mahogany Poly	2864
40	Lime Poly	2866
49	Dark Green Poly	2752
50	Cream	2867
51	Bright Yellow	2094
57	Cream Gold	2884
65	Buckskin	2829
67	Saddle Poly	2871
72	Red	2544
78	Red Orange	2084

In two-tone combinations the first two digits indicate lower color, the next two digits the upper color.

1977 OLDSMOBILE

11	White	2058
19	Ebony Black	9300
13	Silver Poly	2953

16	Med. Gray Poly (Two-Tone)	2954
22	Light Blue Poly	2955
29	Dark Blue Poly	2959
36	Firethorn Red Poly	2811
38	Dark Aqua Poly	2961
44	Medium Green Poly	2964
48	Dark Green Poly	2965
50	Yellow	2884
51	Bright Yellow	2094
61	Light Buckskin	2869
63	Buckskin Poly	2970
69	Brown Poly	2972
72	Red	2973
75	Bright Red	2546
78	Mandarin Orange Poly	2976
85	Med. Blue Poly (Two-Tone)	2980

In two-tone combinations the first two digits indicate lower color, the next two digits the upper color.

1978 OLDSMOBILE

11	White	2058
19	Ebony Black	9300
15	Silver Poly	3076
16	Gray Poly (Accent Color)	3077
21	Pastel Blue	3078
22	Light Blue Poly	2955
24	Bright Blue Poly	3079
29	Dark Blue Poly	2959
32	Drk. Carmel Firemist Poly	3069
33	Lt. Golden Carmel Firemist Poly	3073
44	Light Green Poly	3081
45	Medium Green Poly	3082
48	Dark Green Poly	2965
51	Bright Yellow	3084
56	Medium Gold Poly (Accent Color)	3086
61	Light Camel Beige	3088
63	Medium Camel Poly	3090
67	Russet Poly	3091
69	Dark Camel Poly	3092
75	Bright Red	3095
77	Carmine Red Poly	3096
79	Dark Carmine Poly	3098

In two-tone combinations the first two digits indicate lower color, the next two digits the upper color.

PAINT CODE	COLOR	PPG CODE
1941 PLYMOUTH		
101	Black	9000
201	Aviator Blue Poly	10041
203	Eddins Blue Poly	10050
301	Metallique Green Poly	40080
303	Jib Green No. 5 Poly	40047
305	Hollywood Green	40011
340	Purser Green	40016
401	Plaza Brown No. 3 Poly	20026
410	Casino Beige	20010
501	Airwing Gray	30003
503	Flight Gray Med. Poly	30310
601	Mandarin Maroon	50034

On cars bearing serial numbers higher than Standard 15,008,050 Deluxe 11,149,000 use:

601	Mandarin Maroon	50003
603	Sumach Red	70004
801	Plymouth Gunmetal Medium Poly	30047
803	West Point Gold Poly	20033
805	Charlotte Ivory (To match weathered car)	80039
805	Charlotte Ivory (Original color)	80040
820	Raleigh Tavern Rust	70044
	Aviator Blue Poly (Upper)	10041
901	Eddins Blue Poly (Lower)	10050
	Metallique Green Poly (Upper)	40080
902	Jib Green No. 5 Poly (Lower)	40047
	Airwing Gray (Upper)	30003
	Mandarin Maroon (Lower) (See Paint Code 601)	
	Flight Gray Poly (Upper) (See Paint Code 503)	
904	Plaza Brown No. 3 Poly (Lower)	20026

1942 PLYMOUTH

101	Black	9000
201	Marine Blue	10017
203	Chevron Blue	10018
301	Pilot Green	40012
303	Artillery Green	40007
401	Battalion Beige	20005
501	Airwing Gray	30003
601	Cruiser Maroon	50003
603	Sumach Red	70004
801	Plymouth Gunmetal	30004
803	Charlotte Ivory	80001
	Marine Blue (Upper)	10017
901	Chevron Blue (Lower)	10018
	Chevron Blue (Upper)	10018
902	Marine Blue (Lower)	10017
	Airwing Gray (Upper)	30003
903	Plymouth Gunmetal (Lower)	30004
	Plymouth Gunmetal (Upper)	30004
904	Airwing Gray (Lower)	30003
	Artillery Green (Upper)	40007
905	Pilot Green (Lower)	40012
	Pilot Green (Upper)	40012
906	Artillery Green (Lower)	40007

1946 PLYMOUTH

1	Black	9000
2	Marine Blue	10017
3	Chevron Blue	10319
4	Airwing Gray	30003
5	Cruiser Maroon	50003
6	Balfour Green	40004
7	Kenwood Green	40162
8	Battalion Beige	20005
9	Plymouth Gunmetal	30004
10	Sumach Red	70004
20	Charlotte Ivory	80001

1947 - 1948 PLYMOUTH

All colors used in 1947 - 48 are the same as 1941 to 1946.

1949 PLYMOUTH

601	Black	9000
605	Salvador Blue	10023
606	New Brunswick Blue	10390
620	Kitchener Green	40478
621	Bolivia Green	40170
635	Peru Gray	30368
636	Yukon Gray	30427
645	Edmonton Beige	20396
646	Trinidad Brown	20396
647	Malibu Brown	20209
660	Rio Maroon	50105

661	Mexico Red	70200
665	Plymouth Cream	80380

1950 PLYMOUTH

501	Black	9000
505	Salvador Blue	10023
506	New Brunswick Blue	10390
520	Channel Green	40792
521	Shore Green	40885
535	Peru Gray	30368
536	Gaynor Gray	30571
545	Palm Beige	20679
546	Trinidad Brown	20396
547	Malibu Brown	20209
560	Rio Maroon	50105
561	Mexico Red	70200
565	Plymouth Cream	80469

1951 PLYMOUTH

501	Black	9000
505	Wedgewood Blue 1951	10773
506	New Brunswick Blue	10390
520	Nile Green Poly	41049
520	Nile Green (Solid)	41044
521	Sherwood Green	41075
535	Sterling Gray	30849
536	Luna Gray	30674
545	Palm Beige	20679
560	Mecca Maroon	50212
561	Mexico Red	70200
565	Plymouth Cream	80469

1952 PLYMOUTH

501	Black	9000
505	Wedgewood Blue	10773
507	Belmont Blue Poly	10801
	(Used in first Part of 1952 production)	
507	Coronado Blue Poly	10937
	(Used in latter Part of 1952 production)	
520	Nile Green	41044
522	Lido Green Sympho	41118
535	Sterling Gray	30849
536	Luna Gray	30674
561	Mexico Red	70200
562	Empire Maroon	50248
(Two-Tones only) Dawn Gray		80469

TWO-TONES

670	U 30849	672	U 9000	674	U 31024		
	L 10801		L 41122		L 10937		
670	U 30849	673	U 20823	675	U 31024		

	L 10937		L 20809		L 10773	
671	U 20805	674	U 31024	676	U 31024	
	L 20809		L 10801		L 41118	

1953 PLYMOUTH

601	Black	9000
605	Valencia Blue	10773
607	Coronado Blue Poly	10937
620	Monterey Green	41244
622	Cactus Green Poly	41245
635	Cortez Gray	31010
636	Pecos Gray	31011
645	Suede	20875
646	Sonora Bronze Poly	20876
661	Toreador Red	70200
662	Plaza Maroon Poly	50320
665	Patio Cream	80584

TWO-TONES

670	U 9000	674	U 41245	678	U 31011		
	L 31011		L 41244		L 9000		
671	U 10773	675	U 31010	679	U 20875		
	L 31010		L 10773		L 20876		
672	U 10937	676	U 31010	680	U 20876		
	L 31010		L 10937		L 20875		
673	U 41244	677	U 31010	681	U 50320		
	L 41245		L 50320		L 31010		

1954 PLYMOUTH

601	Black	9000
605	Modesto Blue	11074
606	Avalon Blue Poly	11072
607	San Pedro Blue	11073
615	Berkeley Green	41367
616	Shasta Green Poly	41477
618	San Gabriel Green	41474
630	Pasadena Gray	31127
631	Cascade Gray	31128
640	Pomona Beige	20995
641	Mohave Brown Poly	20997
650	Piedmont Maroon	50364
651	Santa Rosa Coral	70454
655	San Diego Gold	80641
Two Tones/	San Leandro Ivory	80640
Only:	San Mateo Wheat	80073

TWO-TONES

660	U 11072	668	U 31127	676	U 80640		
	L 31127		L 50364		L 41474		
661	U 31127	670	U 9000	677	U 80640		
	L 11072		L 80641		L 70454		

662	U 31127	671	U 9000	678	U 80073
	L 11074		L 11073		L 80641
664	U 20997	672	U 9000	679	U 80073
	L 20995		L 41474		L 11073
665	U 20995	673	U 9000	680	U 80073
	L 20997		L 70454		L 41474
666	U 41367	674	U 80640	681	U 80073
	L 41477		L 80641		L 70454
667	U 41477	675	U 80640		
	L 41367		L 11073		

SPRING COLORS

400	U	Saratoga Ivory	80697
	L	Parakeet Green #2	41579
401	U	Mocha Beige	21097
	L	Oriole Orange	60166
402	U	Solitaire Blue #2	11207
	L	Dutch Blue #3	11172
403	U	Saratoga Ivory	80697
	L	Tinsel Green #1	41595

1955 PLYMOUTH

601	Black	9000
605	Miami Blue	11233
606	Biscayne Blue Poly	11234
607	Tampa Turquoise	11235
615	Tamiami Green	41636
616	Gulf Green Poly	41637
617	Glades Green	41667
618	Largo Green Poly	41677
640	Saratoga Sand	21118
641	Cypress Brown Poly	21113
650	Pompano Peach	60168
651	Seminole Scarlet	70509
655	Orlando Ivory	80697

Two Tones

Only:	Bimini Blue Green Poly	11236

TWO-TONES

661	U 11234	669	U 21113	682	U 9000
	L 11233		L 21118		L 41667
662	U 80697	670	U 21118	683	U 21113
	L 11233		L 21113		L 60168
663	U 11233	671	U 31219	684	U 21118
	L 11234		L 80697		L 60168
664	U 21118	677	U 11236	685	U 9000
	L 11234		L 11235		L 60168
665	U 41637	678	U 21118	686	U 9000
	L 41636		L 11235		L 70509
666	U 41636	679	U 9000	687	U 80697

	L 41636		L 11235		L 70509
667	U 21118	680	U 41677	688	U 80697
	L 41637		L 41667		L 9000
668	U 80697	681	U 21118		
	L 31219		L 41667		

1956 PLYMOUTH

601	Jet Black	9000
605	Powder Blue	11411
607	Wedgewood Blue	11412
608	Turquoise Blue	41763
609	Midnight Blue Poly	11357
615	Sea Spray Green	41841
617	Pine Green Poly	41842
630	Pearl Gray	31320
631	Charcoal Gray Poly	31146
640	Bronze Poly	21209
650	Cherry Red	70621
651	Briar Rose	70598
655	Canary Yellow	80787
657	Eggshell White	8038

Two-Tones

Only:	Pink Champagne	20934
	Cable Red Poly	70722
	Milan Yellow (Green)	42021
	Gazelle Blue	11573
	Salty Green	42020

TWO-TONES

661	U 8038	669	U 8038	677	U 8038
	L 9000		L 41841		L 70621
662	U 8038	670	U 41842	678	U 9000
	L 11411		L 41841		L 70621
663	U 11412	671	U 8038	679	U 8038
	L 11411		L 41842		L 70598
664	U 8038	672	U 8038	680	U 9000
	L 11412		L 31320		L 70598
665	U 8038	673	U 31146	681	U 8038
	L 41763		L 31320		L 80787
667	U 11357	674	U 8038	682	U 9000
	L 41763		L 31146		L 80787
668	U 8038	675	U 8038	683	U 9000
	L 11357		L 21209		L 8038
U & SP.**	8038	U & SP.**	20934		
	L 11573		L 70722		
U & SP.**	8038	U & SP.**	9000		
	L 42020		L 42021		

Note: SP.** Indicates Sportone.

1959 PLYMOUTH

AA	Jet Black	9000
BB	Powder Blue	11796
CC	Starlight Blue Poly	11795
EE	Mint Green	42268
FF	Emerald Green Poly	42270
LL	Pearl Gray	31662
MM	Silver Gray Poly	31663
NN	Flame Red	70813
TT and VV	Bronze Poly	21699
UU	Palomino Beige	21566
WW	Gold Poly	21565
XX	Iceberg White	8131
YY	Daffodil Yellow	80940
ZZ	Bittersweet	60275
RRR	Sunset Beige	70930

NOTE: On two-tone combinations, the first paint code letter indicates the upper color, the second indicates the lower color.

1960 PLYMOUTH

AA-1	Buttercup Yellow	81125
BB-1	Jet Black	9000
CC-1	Sky Blue	11988
DD-1	Twilight Blue Poly	12044
FF-1	Spring Green	42539
GG-1	Chrome Green Poly	42538
JJ-1	Aqua Mist	11787
KK-1	Turquoise Poly	42263
LL-1	Platinum Poly	31746
MM-1	Silver Poly	31661
PP-1	Plum Red Poly	71003
PP-1	Valiant Red	71065
TT-1	Desert Beige	21757
WW-1	Oyster White	8218
YY-1	Caramel Poly	21764

1960 VALIANT SPRING COLORS

CC-1	Valiant Light Blue	11988
FF-1	Valiant Light Green	42539

TWO-TONES: First letter is upper color, second letter (W) is lower, insert color same as lower.
SPORTSTONES: First letter (C) is top color, second letter (W) is lower, insert color same as upper.
SPECIAL SPORTSTONES: First letter (C) is body color, second letter (W) is insert color.

1961 PLYMOUTH

AA-1	Maize	81234
BB-1	Jet Black	9000
CC-1	Robin Egg Blue	12216
DD-1	Airforce Blue Poly	12272
FF-1	Mint Green	42733
GG-1 and NN-1	Emerald Green Poly	42732
KK-1	Twilight Turquoise	12127
LL-1	Silver Gray Poly	31746
PP-1	Carnival Red	71203
RR-1	Coral	71122
SS-1	Lavender Poly (Blue)	12295
WW-1	Alpine White	8218
YY-1	Fawn Beige	21905
ZZ-1	Bronze Poly	21927

TWO-TONES - First letter or digit is accent or roof color. Second letter or digit is basic body color.

1962 PLYMOUTH

AA-1	Sun-Glo	81285
BB-1	Silhouette Black	9000
CC-1	Pale Blue	12403
DD-1	Luminous Blue Poly	12416
FF-1	Pale Jade	42840
GG-1	Luminous Green Poly	42854
KK-1	Luminous Turquoise Poly	12259
MM-1	Light Gray	32127
PP-1	Cherry Red	71203
SS-1	Luminous Cordovan Poly	22025
TT-1	Sandstone	32074
WW-1	Ermine White	8293
YY-1	Luminous Brown Poly	22023

TWO-TONES: First letter or digit is accent or roof color. Second letter or digit is basic body color.

1963 PLYMOUTH

BB-1	Ebony Black	9000
CC-1	Light Blue	12469
DD-1	Medium Metallic Blue Poly	12625
EE-1	Dark Metallic Blue Poly	12658
FF-1	Light Green	43081
GG-1	Metallic Green Poly	43055
MM-1	Light Beige	32202
NN-1	Valiant Silver Gray Poly	32263
PP-1	Red	71203
RR-1	Coppertone Poly	22240
WW-1	Ermine	8293
XX-1	Medium Beige	22234
YY-1	Metallic Brown Poly	22218

TWO-TONES: First letter or digit is accent or roof color. Second letter or digit is basic body color.

1964 PLYMOUTH

AA-1	Barracuda Gold Poly	22461
BB-1	Ebony Black	9000
CC-1	Light Blue	12655
DD-1	Medium Blue Poly	12656
EE-1	Dark Blue Poly	12657
HH-1	Sandalwood Poly	22237
JJ-1	Valiant Light Turquoise	12736
KK-1	Turquoise Poly	12647
LL-1	Dark Turquoise Poly	12708
MM-1	Silver Gray Poly	32263
PP-1	Ruby	71203
TT-1	Signet Royal Red Poly	71355
VV-1	Chestnut Poly	22238
WW-1	White	8358
XX-1	Light Beige	22293
YY-1	Medium Beige Poly	22296

TWO-TONES: First letter or digit is accent or roof color. Second letter or digit is basic body color.

1965 PLYMOUTH

AA-1	Gold Poly	22461
BB-1	Black	9000,9300
CC-1	Light Blue	12894
DD-1	Medium Blue Poly	12763
EE-1	Dark Blue Poly	12896
HH-1	Copper Poly	22445
JJ-1	Light Turquoise	12901
KK-1	Medium Turquoise Poly	12897
LL-1	Dark Turquoise Poly	12765
NN-1	Barracuda Silver Poly	32398
PP-1	Ruby	71433
SS-1	Ivory	81413
TT-1	Medium Red Poly	71476
WW-1	White	8362
XX-1	Light Tan	22440
YY-1	Medium Tan Poly	22643

TWO-TONES: First letter or digit is accent or roof color. Second letter or digit is basic body color.

1966 PLYMOUTH

AA-1	Silver Poly	32398
BB-1	Black	9000,9300
CC-1	Light Blue	13037
DD-1	Light Blue Poly	13043
EE-1	Dark Blue Poly	13040
GG-1	Dark Green Poly	43149
KK-1	Light Turquoise Poly	12898
LL-1	Dark Turquoise Poly	12765

MM-1	Turbine Bronze Poly	60492
PP-1	Bright Red	71483
QQ-1	Dark Red Poly	71476
RR-1	Yellow	81515
SS-1	Soft Yellow	81501
WW-1	White	8362
XX-1	Beige	22541
YY-1	Bronze Poly	22538
ZZ-1	Citron Gold Poly	22511
22-1	Charcoal Poly	32513
66-1	Light Mauve Poly	50702

TWO-TONES: First letter or digit is accent or roof color. Second letter or digit is basic body color.

1967 PLYMOUTH

AA-1	Silver Poly	32398
BB-1	Black	9300,9000
CC-1	Medium Blue Poly	13159
DD-1	Light Blue Poly	13043
EE-1	Dark Blue Poly	13040
FF-1	Light Green Poly	43547
GG-1	Dark Green Poly	43540
HH-1	Dark Copper Poly	22659
KK-1	Light Turquoise Poly	13195
LL-1	Dark Turquoise Poly	13214
MM-1	Turbine Bronze Poly	60492
PP-1	Bright Red	71483
QQ-1	Dark Red Poly	71552
RR-1	Yellow	81515
SS-1	Soft Yellow	81501
TT-1	Medium Copper Poly	22706
WW-1	White	8362
XX-1	Beige	22701
YY-1	Light Tan Poly	22700
ZZ-1	Gold Poly	22715
66-1	Mauve Poly	50731
88-1	Bright Blue Poly	13336

TWO-TONES: First letter or digit is accent or roof. Second letter or digit is basic color.

1968 PLYMOUTH

AA-1	Buffed Silver Poly	8588
BB-1	Black	9000,9300
CC-1	Medium Blue Poly	13355
DD-1	Mist Blue Poly	13360
EE-1	Midnight Blue Poly	13372
FF-1	Mist Green Poly	43646
GG-1	Forest Green Poly	43649
HH-1	Yellow Gold	81575

JJ-1	Ember Gold Poly	22807
KK-1	Mist Turquoise Poly	13195
LL-1	Surf Turquoise Poly	13371
MM-1	Turbine Bronze Poly	60492
PP-1	Matador Red	71483
QQ-1	Electric Blue Poly	13354
RR-1	Burgundy Poly	50749
SS-1	Sunfire Yellow	81574
TT-1	Avocado Green Poly	43647
UU-1	Frost Blue Poly	13445
WW-1	Sable White	8653
XX-1	Satin Beige	22441
YY-1	Sierra Tan Poly	22855
22-1	Hawaiian Blue	13631

TWO-TONES: First letter or digit is accent or roof color. Second letter or digit is basic body color.

1969 PLYMOUTH

A-4	Silver Poly	2016
B-3	Ice Blue Poly	2018
B-5	Blue Fire Poly	2019
B-7	Jamaica Blue Poly	2020
F-3	Frost Green Poly	2023
F-5	Limelight Poly	2024
F-8	Ivy Green Poly	43786
X-9	Black Velvet	9300
K-2	Vitamin "C" Orange	2201
L-1	Sandpebble Beige	22542
Q-5	Seafoam Turquoise Poly	13534
R-4	Barracuda Orange	71582
R-6	Scorch Red	2029
T-3	Honey Bronze Poly	2030
T-5	Bronze Fire Poly	2031
T-7	Saddle Bronze Poly	2032
W-1	Alpine White	2033
Y-2	Sunfire Yellow	81574
Y-3	Yellow Gold	81575
Y-4	Spanish Gold Poly	2034
Y-6	Citron Gold Poly	2102
999	Orange	60436
999	Rallye Green	44032
999	Bahama Yellow	81570

TWO-TONES: First letter or digit is accent or roof color. Second letter or digit is basic body color.

1970 PLYMOUTH

A-4	Silver Poly	2016
B-3	Ice Blue Poly	2018
B-5	Blue Fire Poly	2019
B-7	Jamaica Blue Poly	2020
C-7	In Violet	2210
E-5	Rallye Red	2136
F-4	Lime Green Poly	2133
F-8	Ivy Green Poly	43786
J-5	Limelight Poly	2128
X-9	Black Velvet	9300
J-6	Sassy Grass Green	2259
K-2	Vitamin "C"	2201
K-3	Burnt Orange Poly	2134
K-5	Deep Burnt Orange Poly	2135
L-1	Sandpebble Beige	22542
M-3	Moulin Rouge	2260
P-6	Frosted Teal Poly	2132
R-6	Scorch Red	2029
T-3	Sahara Tan Poly	2131
T-6	Burnt Tan Poly	2129
T-8	Walnut Poly	2130
V-2	Tor-Red	2186
W-1	Alpine White	2033
Y-1	Lemon Twist	2211
Y-2	Sunfire Yellow	81574
Y-3	Yellow Gold	81575
Y-4	Citron Mist Poly	2117
Y-6	Citron Gold Poly	2102

TWO-TONES: First letter or digit is accent or roof color. Second letter or digit is basic body color.

1971 PLYMOUTH

A-4	Winchester Gray Poly	2314
A-8	Slate Gray Poly	2315
B-2	Glacial Blue Poly	2304
B-5	True Blue Poly	2306
B-7	Evening Blue Poly	2302
C-7	In-Violet Poly	2210
C-8	Mood Indigo Poly	2305
E-5	Rallye Red	2136
E-7	Burnished Red Poly	2321
F-3	Amber Sherwood Poly	2316
F-7	Sherwood Green Poly	2317
J-4	April Green Poly	2319
J-6	Sassy-Grass Green	2259
K-6	Autumn Bronze Poly	2312
L-1	Sandalwood Beige	22542
L-5	Bahama Yellow	2325
Q-5	Coral Turquoise Poly	2301
T-2	Tunisian Tan Poly	2313
T-8	Tahitian Walnut Poly	2309
V-2	Tor-Red	2186
W-1	Spinnaker White	2033

W-3	Sno-White	2300
X-9	Formal Black	9300
Y-3	Curious Yellow	2320
Y-4	Light Gold	2310
Y-8	Gold Leaf Poly	2307
Y-9	Tawny Gold Poly	2311

TWO-TONES: First letter or digit is accent or roof color. Second letter or digit is basic body color.

1972 PLYMOUTH

A-4	Winchester Gray Poly	2314
A-5	Silver Frost Poly	2513
A-9	Charcoal Poly	2017
B-1	Blue Sky	2424
B-3	Basin Street Blue	2423
B-5	True Blue Poly	2306
B-7	Evening Blue Poly	2302
B-9	Regal Blue Poly	2508
DT-7596	Bright Green	2224
E-5	Red	2136
F-1	Mist Green	2515
F-3	Amber Sherwood Poly	2316
F-7	Sherwood Green Poly	2317
F-8	Forest Green Poly	2514
J-3	Meadow Green	2383
JY-9	Tahitian Gold Poly	2510
L-4	Sahara Beige	2427
Q-5	Coral Turquoise Poly	2301
T-6	Mojave Tan Poly	2426
T-8	Chestnut Poly	2425
V-2	Tor-Red	2186
W-1	Spinnaker White	2033
Y-1	Lemon Twist	2211
Y-2	Sun Fire Yellow	81574
Y-3	Honey Gold	2517
Y-4	Honeydew	2310
Y-6	Golden Haze Poly	2509
Y-8	Gold Leaf Poly	2307
Y-9,(HY-9)	Tawney Gold Poly	2311
X-9	Formal Black	9300

TWO-TONES: First letter or digit is accent or roof color. Second letter or digit is basic body color.

1973 PLYMOUTH

A-5	Silver Frost Poly	2513
B-1	Blue Sky	2424
B-3	Basin Street Blue	2423
B-5	True Blue Poly	2306
B-9	Regal Blue	2508

E-5	Rallye Red	2136
F-1	Mist Green	2515
F-3	Amber Sherwood Poly	2316
F-8	Forest Green Poly	2514
K-6	Autumn Bronze Poly	2312
L-4	Sahara Beige	2427
Q-5	Coral Turquoise Poly	2301
T-6	Mojave Tan Poly	2426
T-8	Chestnut Poly	2425
T-9	Chestnut Poly	2590
W-1	Spinnaker White	2033
Y-1	Lemon Twist	2211
Y-2	Sunfire Yellow	81574
Y-3	Honey Gold	2517
Y-6	Golden Haze Poly	2509
JY-9	Tahitian Gold Poly	2510
X-9	Formal Black	9000,9300

TWO-TONES: First letter or digit is accent or roof color. Second letter or digit is basic body color.

1974 PLYMOUTH

A-5	Silver Frost Poly	2513
B-1	Powder Blue	2626
B-5	Lucerne Blue Poly	2627
B-8	Starlight Blue Poly	2628
E-5	Rallye Red	2136
E-7	Burnished Red Poly	2321
G-2	Frosty Green Poly	2629
G-8	Deep Sherwood Poly	2631
J-6	Avocado Gold Poly	2632
L-4	Sahara Beige	2427
L-6	Aztec Gold Poly	2591
L-8	Dark Moonstone Poly	2633
T-5	Sienna Poly	2634
T-9	Dark Chestnut Poly	2590
W-1	Spinnaker White	2033
Y-2	Sunfire Yellow	81574
Y-4	Golden Fawn	2635
Y-5	Yellow Blaze	2636
Y-6	Golden Haze Poly	2509
Y-9	Tahitian Gold Poly	2510
X-9	Formal Black	9300

TWO-TONES: First letter or digit is accent or roof color. Second letter or digit is basic body color.

1975 PLYMOUTH

A-2	Silver Cloud Poly	2734
B-1	Powder Blue	2626
B-2	Astral Blue Poly	2735

B-5	Lucerne Blue Poly	2627
B-8	Starlight Blue Poly	2628
E-5	Rallye Red	2136
E-9	Vintage Red Poly	2736
G-2	Frosty Green Poly	2629
G-8	Deep Sherwood Poly	2631
J-2	Platinum Poly	2730
J-6	Avocado Gold Poly	2632
K-3	Bittersweet Poly	2740
L-4	Sahara Beige	2427
L-5	Moondust Poly	2737
L-6	Aztec Gold Poly	2591
T-4	Cinnamon Poly	2741
T-5	Sienna Poly	2634
T-9	Dark Chestnut Poly	2590
W-1	Spinnaker White	2033
Y-4	Golden Fawn	2635
Y-5	Yellow Blaze	2636
Y-6	Inca Gold Poly	2738
Y-9	Spanish Gold Poly	2739
X-9	Formal Black	9300

TWO-TONES: First letter or digit is accent or roof color. Second letter or digit is basic body color.

1976 PLYMOUTH

A-2	Silver Cloud Poly	2734
B-1	Powder Blue	2626
B-2	Astral Blue Poly	2735
B-4	Big Sky Blue	2850
B-5	Jamaican Blue Poly	2851
B-8	Starlight Blue Poly	2628
E-5	Rallye Red	2136
E-9	Vintage Red Poly	2736
F-2	Jade Green Poly	2852
G-8	Deep Sherwood Poly	2631
J-2	Platinum Poly	2730
J-5	Tropic Green Poly	2853
K-3	Bittersweet Poly	2740
L-4	Sahara Beige	2427
L-5	Moondust Poly	2737
R-6	Claret Red	2854
T-4	Cinnamon Poly	2741
T-9	Dark Chestnut Poly	2590
U-2	Saddle Tan	2855
U-3	Carmel Tan Poly	2856
V-1	Spitfire Orange	2858
W-1	Spinnaker White	2033
Y-3	Harvest Gold	2859
Y-4	Golden Fawn	2635
Y-5	Yellow Blaze	2636

Y-6	Inca Gold Poly	2738
Y-7	Taxi Yellow	81746
Y-9	Spanish Gold Poly	2739
X-9	Formal Black	9000,9300

TWO-TONES: First letter or digit is accent or roof color. Second letter or digit is basic body color.

1977 PLYMOUTH

A-2	Silver Cloud Poly	2734
B-2	Wedgewood Blue	2934
B-3	Cadet Blue Poly	2935
B-5	French Racing Blue	2936
B-6	Regatta Blue Poly	2937
B-9	Starlight Blue "Sunfire" Poly	2938
E-5	Rallye Red	2136
E-8	Vin. Red "Sunfire" Poly	2849
F-2	Jade Green Poly	2852
F-7	For. Grn. "Sunfire" Poly	2939
K-6	Burn. Copper Poly	2940
L-3	Mojave Beige	2941
L-5	Moondust Poly	2737
R-8	Russet "Sunfire" Poly	2942
T-2	Light Mocha Tan	2943
T-7	Coffee "Sunfire" Poly	2944
U-3	Caramel Tan Poly	2856
V-1	Spitfire Orange	2858
W-1	Spinnaker White	2033
X-8	For. Blk. "Sunfire" Poly	2945
Y-1	Jasmine Yellow	2946
Y-3	Har. Gold (Two-Tone)	2859
Y-4	Golden Fawn	2635
Y-5	Yellow Blaze	2636
Y-6	Inca Gold Poly	2738
Y-7	Taxi Yellow	81746
Y-9	Spanish Gold Poly	2739

TWO-TONES: First letter or digit is accent or roof color. Second letter or digit is basic body color

1978 PLYMOUTH

EW-1	Spinnaker White	2033
EY-7	Taxi Yellow	81746
JA-5	Silver Frost (Two-Tone)	2513
KY-4	Golden Fawn	2635
KY-5	Yellow Blaze	2636
LY-9	Spanish Gold Poly	2739
MU-3	Caramel Tan Poly	2856
MV-1	Spitfire Orange	2858
PB-2	Wedgewood Blue	2934
PB-3	Cadet Blue Poly	2935

PB-6	Regatta Blue Poly	2937
PB-9	Star. Bl. "Sunfire" Poly	2938
PT-2	Light Mocha Tan	2943
PY-1	Jasmine Yellow	2946
RA-2	Pewter Grey Poly	3016
RA-9	Char. Grey "Sunfire" Poly	3017
RF-3	Mint Green Poly	3019
RF-9	Aug. Grn. "Sunfire" Poly	3018
RJ-3	Citron Poly	3066
RK-2	Sunrise Orange	3067

RR-4	Brite Canyon Red	3068
RR-7	Tap. Red "Sunfire" Poly	3020
RR-9	Crim. Red "Sunfire" Poly	3064
RT-9	Sable "Sunfire" Poly	3014
RY-3	Classic Cream	3021
TX-9	Black	9000,9300

TWO-TONES: First letter or digit is accent or roof color.
Second letter or digit is basic body color.

Chapter 14   PONTIAC 1941-1978

PAINT CODE	COLOR	PPG CODE
1941 PONTIAC		
4100	Black	9200
4101	Marlboro Blue Poly	10080
4102	Parma Wine Poly	50014
4104	Streamline Gray Poly	30055
4105	Tropic Blue Poly	10079
4106	Allandale Green Poly	40125
4107	Indiana Beige Poly	20038
4108	Taffy Tan	20082
4109	El Paso Beige Poly (Upper)	20037
	Indiana Beige Poly (Lower)	20038
4110	Paddock Gray Poly (Upper)	30112
	Marlboro Blue Poly (Lower)	10080
4111	Silver French Gray Poly (Upper)	30051
	Streamline Gray Poly (Lower)	30055
4112	Thetis Green Poly (Upper)	40124
	Allandale Green Poly (Lower)	40125
4114	French Copper (Upper)	20083
	Taffy Tan (Lower)	20082
4115	Santone Beige Poly (Upper)	20081
	Parma Wine Poly (Lower)	50014
4116	Taffy Tan (Upper)	20082
	French Copper (Lower)	20083
4117	Pearl Gray Poly (Upper)	30071
	Black (Lower)	9001
1942 PONTIAC		
4200	Black	9200
4201	Regency Blue Poly	10104
4202	Parma Wine Poly	50014
4204	Beverly Gray Poly	30087
4205	Skipper Blue Poly	10105
4206	Milori Green	40108
4207	Beaver Head Brown Poly	20063
4209	Mermaid Green	40088
4210	Pearl Gray Poly (Upper)	30094
	Black (Lower)	9001
4211	Pearl Gray Poly (Upper)	30094
	Regency Blue Poly (Lower)	10104
4214	Lucian Gray (Upper)	30088
	Beverly Gray Poly (Lower)	30087
4215	Casino Gray (Upper)	30089
	Skipper Blue Poly (Lower)	10105
4216	Harbor Gray (Upper)	30090
	Milori Green (Lower)	40108
4217	Rosewood Beige (Upper)	20064

PAINT CODE	COLOR	PPG CODE
	Beaver Head Brown Poly (Lower)	20063
4218	Radnor Beige (Upper)	20065
	Cinnebar Red (Lower)	70031
4219	Caliente Green (Upper)	40102
	Mermaid Green (Lower)	40088
4208	Cinnebar Red	70031
1946 PONTIAC		
4600	Black	9200
4601	Mariner Blue Poly	10312
4602	Parma Wine Poly	50014
4604	Silverwing Gray Poly	30322
4607	Smoked Pearl Poly	30321
4614	Mariner Blue Poly (Upper)	10312
	Silverwing Gray Poly (Lower)	30322
4617	Silverwing Gray Poly (Upper)	30322
	Smoked Pearl Poly (Lower)	30321
1947 PONTIAC		
4700	Black	9200
4701	Mariner Blue Poly	10312
4702	Parma Wine Poly	50014
4704	Silverwing Gray Poly	30322
4705	Asbury Green Poly	40443
4707	Smoked Pearl Poly	30321
4708	Cairo Cream	80118
	(Two-Tone only) Burbank Green Poly	40444

TWO-TONES

4714	U 10312	4715	U 40444	4717	U 30322
	L 30322		L 40443		L 30321

PAINT CODE	COLOR	PPG CODE
1948 PONTIAC		
	Black	9200
4801	Mariner Blue Poly	10312
4802	Parma Wine Poly	50125
4804	Belgian Gray Poly	30512
4805	Frances Ivory	80349
4806	Rio Red	50124
4807	Genesee Green Poly	40567
4809	Oyster Gray	30511

TWO-TONES

4814	U 30511	4818	U 30511
	L 30512		L 10468
4817	U 40103	4819	U 10468
	L 40567		L 30511

149

1949 PONTIAC

	Black	9200
901	Starlight Blue Poly	10561
903	Parma Wine Poly	50125
904	Sheffield Gray Poly	30626
905	Nankeen Cream	80400
906	Rio Red	50124
907	Sage Green	40705
908	Blue Lake Blue Poly	10468
909	Coventry Gray	30627
910	Wellington Green Poly	40706
911	Mayan Gold Poly	20564

TWO-TONES

4914	U 30626	4918	U 30627
	L 30627		L 10468
4917	U 40706	4919	U 10468
	L 40705		L 30627

1950 PONTIAC

000	Black	9200
001	Starlight Blue Poly	10561
003	Parma Wine Poly	50125
004	Warwick Gray Poly	30744
005	San Pedro Ivory	80474
006	Rio Red	50124
007	Tarragon Green	40914
008	Skylark Blue Poly	10677
009	Cavalier Gray	30743
010	Berkshire Green Poly	40913
011	Solar Gold Poly	20684
012	Sierra Rust	20683
024	Lido Beige	20696
028	San Leandro Cream	80473

TWO-TONES

014	U 30744	5018	U 30743	5023	U 40913
	L 30743		L 10677		L 40914
015	U 20683	5019	U 10677	5024	U 20684
	L 80474		L 30743		L 20696
017	U 40913	5022	U 80474	5028	U 20683
	L 40914		L 20683		L 80473

1951 PONTIAC

100	Black	9200
101	St. Clair Blue Poly	10784
103	Victoria Maroon Poly	50229
104	Yorkshire Gray Poly	30879
105	Malibu Ivory	20781
106	Tripoli Red	70368
107	Palmetto Green	41078
108	Starmist Blue	10785
109	Surf Gray	30878
110	Berkshire Green Poly	40913
111	Saturn Gold Sympho	20780
112	Sapphire Poly	10795
112	Sapphire Poly	10786
123	Berkshire Green Poly	40913
124	Lido Beige	20696
124	Saturn Gold Sympho	20780
125A	Imperial Maroon	50231
125B	Sand Gray Poly	30880
5126	Berkshire Green Poly	40913
5126	Palmetto Green	41078

TWO-TONES

5114	U 30879	5117	U 40913	5122	U 20781
	L 30878		L 41078		L 10795
5115	U 10786	5119	U 10785	5128	U 10784
	L 10795		L 30878		L 10785
5115	U 10795	5122	U 20781		
	L 20781		L 10786		

1952 PONTIAC

5200	Black	9200
5201	Potomac Blue Poly	10829
5203	Victoria Maroon Poly	50229
5204	Smoke Gray Poly	30935
5205	Seamist Green	41144
5206	Cherokee Red	70384
5207	Placid Green	40561
5208	Mayflower Blue Poly	10830
5209	Shell Gray	30936
5210	Forest Green Poly	41146
5211	Saturn Gold Sympho	20780
5212	Belfast Green	41145
5224	Lido Beige	20696
5225	Imperial Maroon	50231
5225	Sand Gray Poly	30880

TWO-TONES PASSENGER CARS

5214	U 30935	5217	U 41146	5222	U 41144
	L 30936		L 40561		L 41145
5215	U 411455	219	U 10830	5228	U 10829
	L 41144		L 30936		L 10830

TWO-TONES STATION WAGON

5223	(BC) 41146	5225	(BC) 50231
	(PI) 40561		(PI) 30880
5224	(BC) 20780	5228	(BC) 40561
	(PI) 20696		(PI) 41146

BC = Body Color   PI = Panel Insert

1953 PONTIAC

5300	Black	9200
5301	Caravan Blue Poly	10940
5303	Continental Maroon	50331
5304	Marathon Gray Poly	31054
5305	Milano Ivory	80602
5306	Santa Fe Red	70414
5307	Linden Green	41306
5308	Stardust Blue	10961
5309	Cirro Gray	31055
5310	Spruce Green Poly	41305
5311	Winona Green Poly	41343
5312	Laurel Green	41304

TWO-TONES

5314	U 31054 L 31055	5317	U 41305 L 41306	5319	U 10961 L 31055
5315	U 41304 L 80602	5318	U 10940 L 10961	5322	U 80602 L 41304

1954 PONTIAC

5400	Raven Black	9200
5401	San Marino Blue Poly	11155
5402	Seneca Brown Poly	21071
5403	Arlington Maroon	50381
5404	Cruiser Gray Poly	31188
5405	Maize Yellow	80675
5406	Picador Red	70481
5407	Shannon Green	41575
5408	Mayfair Blue	11154
5409	Cirro Gray	31055
5410	Brookmere Green Poly	41576
5411	Coral Red	70488
5412	Biloxi Beige	21070

Two Tones

Only	Winter White	80674

TWO-TONES

5413	U 9200 L 41575	5417	U 41576 L 41575	5421	U 80674 L 70488
5414	U 31188 L 31055	5418	U 11155 L 11154	5422	U 80674 L 21070
5415	U 80674 L 80675	5419	U 11154 L 31055		
5416	U 9200 L 70481	5420	U 9200 L 31055		

1955 PONTIAC

5500	Raven Black	9200

5501	Beaumont Blue Poly	11331
5502	Corsair Tan	21152
5503	Persian Maroon Poly	50406
5504	Falcon Gray Poly	31272
5505	Avalon Yellow	80668
5506	Bolero Red	70576
5507	Valley Green	41741
5508	Marietta Blue	11148
5509	Castle Gray	31271
5510	Sequoia Green Poly	41742
5511	Firegold Poly	21153
5512	Turquoise Blue Poly	11330
5531	Nautilus Blue	11326
5541	Driftwood Beige	21062
5582	Polo White	8102

Two Tones

Only	White Mist	80741

TWO-TONES

Conventional Model	Vogue Model	
5513	5553	U 9200 L 41741
5514	5554	U 31272 L 31271
5515	5555	U 9200 L 80668
5516	5556	U 9200 L 70576
5517	5557	U 41742 L 41741
5518	5558	U 11331 L 11148
5519	5559	U 11148 L 31271
5520	5560	U 9200 L 31271
5521	5561	U 80741 L 21153
5522	5562	U 80741 L 11330
5523	5563	U 31271 L 21152
5524	5564	U 31271 L 11326
5525	5565	U 41742 L 80668
5526	5566	U 11326 L 31271
5527	5567	U 80668 L 9200

5528	5568	U 9200
		L 11326
5529	5569	U 80668
		L 41742
5530	5570	U 11326
		L 9200
5532	5572	U 9200
		L 41741
5533	5573	U 31271
		L 31272
5534	5574	U 70576
		L 9200
5535	5575	U 41741
		L 41742
5536	5576	U 11148
		L 11331
5537	5577	U 31271
		L 11148
5538	5578	U 31271
		L 9200
5539	5579	U 21152
		L 31271
	5582	U 8102
		L 21062

1956 PONTIAC

6A	Raven Black	9200
6B	Chesapeake Blue Poly	11497
6C	Olympic Blue	11498
6D	Amethyst Poly	50441
6E	Phantom Gray Poly	31374
6F	Grenada Gold	21260
6G	Bolero Red	70674
6H	Hialeah Green	41936
6J	Vista Blue	11180
6K	Nimbus Gray	31373
6L	Glendale Green Poly	41935
6M	Terragon Green	40914
6N	Sandalwood Tan Poly	21259
6P	Sun Beige	21258
6Q	Catalina Blue Poly	11496
6R	Camellia	70675
&W	Sun Beige	21330

Note: Colors used in Two-Tones can be Identified by the paint code letters, such as B = Chesapeake Blue Poly., C = Olympic Blue, etc.

1957 PONTIAC

57-A	Raven Black	9000
57-B	Chevron Blue Poly	11622
57-C	Nassau Green	42059
57-D	Sapphire Blue	11326
57-E	Chateau Gray Poly	31450
57-F	Fontaine Blue Poly	11618
57-G	Tartan Red	70740
57-H	Charcoal Gray Poly	31451
57-J	Lucerne Blue	11604
57-K	Sheffield Gray	31446
57-L	Braeburn Green Poly	42072
57-M	Starlight Yellow	80902
57-N	Cordova Red Poly	50468
57-P	Kenya Ivory	80903
57-Q	Silver Gray Poly	31449
57-R	Carib Coral	70741
57-S	Limefire Green Poly	42060
57-T	Seacrest Green	42071
57-U	Sage Blue Poly	11619
56-I	Malabar Yellow	80873
56-O	Kerry Green Poly	42022
56-U	Marina Blue	11575
56-V	Avalon Blue Poly	11576
56-X	Rodeo Beige	21329
56-Z	Lilac Poly	50464

NOTE: Single-tones are designated by double paint code letters, such as BB, CC, etc. Colors used in two-tones can also be identified by the paint code letters, such as (BC) B = Chevron Blue Poly, C = Nassau Green, etc.

1958 PONTIAC

AAA	Persian Black	9000
BBB	Ascot Gray Poly	31623
DDD	Squadron Blue Poly	11749
EEE	Viking Blue	11748
FFF	Darby Green Poly	42217
GGG	Seaforth Green	42220
HHH	Rangoon Red	70823
JJJ	Sunmist Yellow	80960
KKK	Reefshell Pink	70822
LLL	Tropicana Turquoise	42221
MMM	Lilac Mist Poly	50500
NNN	Mallard Turquoise Poly	11751
PPP	Marlin Turquoise	42219
QQQ	Deauville Blue Poly	11747
RRR	Kashmir Blue	11750
SSS	Burma Green Poly	42222
TTT	Calypso Green	42218
UUU	Redwood Copper Poly	21516
VVV	Patina Ivory	8155
WWW	Starmist Silver Poly	31624
YYY	Graystone White	8156
ZZZ	Jubilee Gold Poly	21576

57-V	Cascade Blue	11685
57-W	Mayfair Yellow	80938
57-X	Iris	50484

NOTE: On two-tones and insert combinations the first letter indicates color used on lower area, the second upper area and the third letter indicates color used on insert area.

1959 PONTIAC

AA	Regent Black	9300
BB	Silvermist Gray Poly	31827
CC	Cameo Ivory	8160
DD	Dundee Green Poly	42479
EE	Jademist Green Poly	42480
FF	Seaspray Green	42499
HH	Vanguard Blue Poly	12002
JJ	Castle Blue	12003
KK	Gulfstream Blue Poly	12001
LL	Mandalay Red	70961
MM	Sunset Glow Poly	70959
NN	Royal Amethyst Poly	50536
PP	Shoreline Gold Poly	21722
RR	Canyon Copper Poly	21723
SS	Concord Blue Poly	12025
58-C	Palomar Yellow	80986
58-I	Frontier Beige	21710
58-O	Sunrise Coral	60323
58-X	Orchid	50503

TWO-TONES: The first letter indicates the lower color, the second letter indicates the upper color.

1960 PONTIAC

A	Regent Black	9300
B	Black Pearl Poly	31905
C	Shelltone Ivory	8259
D	Richmond Gray Poly	31928
F	Newport Blue Poly	12174
H	Skymist Blue Poly	12234
J	Fairway Green Poly	42650
K	Berkshire Green Poly	42693
L	Coronado Red Poly	50568
M	Stardust Yellow	81202
N	Mahogany Poly	21874
P	Shoreline Gold Poly	21722
R	Palomino Beige	21873
S	Sierra Copper Poly	21841
T	Caribe Turquoise Poly	12228

Two-tone combinations can be identified by the paint code letters.

TWO-TONES: The first letter indicates the lower color, the second letter indicates the upper color.

1961 PONTIAC

61-A	Regent Black	9300
61-C	Shelltone Ivory	8259
61-D	Richmond Gray Poly	31928
61-E	Bristol Blue Poly	12397
61-F	Richelieu Blue Poly	12398
61-H	Tradewind Blue	12399
61-J	Jadestone Green Poly	42837
61-K	Seacrest Green	42838
61-L	Coronado Red Poly	50568
61-M	Bamboo Cream	81271
61-N	Cherrywood Bronze Poly	21874
61-P	Rainier Turquoise Poly	12396
61-R	Fernando Beige	21733
61-S	Dawnfire Mist Poly	71211
61-T	Mayan Gold Poly	22005

TWO-TONES: The first letter indicates the lower color, the second letter indicates the upper color.

1962 PONTIAC

A	Starlight Black	9300
C	Cameo Ivory	8259
D	Silvermist Gray Poly	32173
E	Ensign Blue Poly	12552
F	Yorktown Blue Poly	12546
H	Kimberley Blue	12549
J	Silverleaf Green Poly	42975
L	Belmar Red Poly	50568
M	Bamboo Cream	81271
N	Burgundy Poly	50615
P	Aquamarine Poly	12525
Q	Seafoam Aqua	12550
R	Yuma Beige	22137
T	Caravan Gold Poly	22121
V	Mandalay Red	70961

TWO-TONES: The first letter indicates the lower color, the second letter indicates the upper color.

1963 PONTIAC

A	Starlight Black	9300
C	Cameo Ivory	8259
D	Silvermist Gray Poly	32173
F	Yorktown Blue Poly	12546
H	Kimberley Blue	12713
J	Silverleaf Green Poly	42975
K	Cordovan Poly	22294
L	Marimba Red Poly	50633

P	Aquamarine Poly	12525
Q	Marlin Aqua Poly	43114
R	Yuma Beige	22137
S	Saddle Bronze Poly	22269
T	Caravan Gold Poly	22268
V	Grenadier Red	71336
W	Nocturne Blue Poly	12696

TWO-TONES: The first letter indicates the lower color, the second letter indicates the upper color.

1964 PONTIAC

A	Starlight Black	9300
C	Cameo Ivory	8259
D	Silvermist Gray Poly	32173
F	Yorktown Blue Poly	12546
H	Skyline Blue	12847
J	Pinehurst Green Poly	43264
L	Marimba Red Poly	50633
L	Marimba Red Poly #2	50684
N	Sunfire Red Poly	71415
P	Aquamarine Poly	12525
Q	Gulfstream Aqua Poly	12848
R	Alamo Beige	22391
S	Saddle Bronze Poly	22269
T	Singapore Gold Poly	22392
V	Grenadier Red	71336
W	Nocturne Blue Poly	12696

TWO-TONES: The first letter indicates the lower color, the second letter indicates the upper color.

1965 PONTIAC

A	Starlight Black	9300
B	Blue Charcoal Poly	32448
C	Cameo Ivory	8259
D	Fontaine Blue Poly	13042
E	Nightwatch Blue Poly	13002
H	Palmetto Green Poly	43391
K	Reef Turquoise Poly	43364
L	Teal Turquoise Poly	13003
N	Burgundy Red Poly	50700
P	Iris Mist Poly	50693
R	Montero Red	71472
T	Capri Gold Poly	22564
V	Mission Beige	22270
W	Bluemist Slate Poly	32461
Y	Mayfair Maize	81500
	Tiger Gold Poly	22728

TWO-TONES: The first letter indicates the lower color, the second letter indicates the upper color.

1966 PONTIAC

A	Starlight Black	9300
B	Blue Charcoal Poly	32448
C	Cameo Ivory	8259
D	Fontaine Blue Poly	13042
E	Nightwatch Blue Poly	13002
F	Cadet Blue Poly	13148
H	Palmetto Green Poly	43391
J	Pinehurst Green Poly	43390
K	Reef Turquoise Poly	43364
L	Marina Turquoise Poly	43496
M	Sierra Red Poly	71525
N	Burgundy Poly	50700
P	Barrier Blue Poly	13242
P	Plum Mist Poly	50717
R	Montero Red	71472
T	Martinique Bronze Poly	22660
V	Mission Beige	22270
W	Platinum Poly	32525
Y	Candlelight Cream	81528
-	Fathom Turquoise	13331
-	Tiger Gold Poly	22728
-	Copper Blaze Poly	60497
-	Ramada Bronze Poly	22803

TWO-TONES: The first letter indicates the lower color, the second letter indicates the upper color.

1967 PONTIAC

A	Starlight Black	9300
C	Cameo Ivory	8259
D	Montreux Blue Poly	13349
E	Fathom Blue Poly	13346
F	Tyrol Blue Poly	13364
G	Signet Gold Poly	22818
H	Linden Green Poly	43651
K	Gulf Turquoise Poly	43661
L	Mariner Turquoise Poly	43659
M	Plum Mist Poly	50717
N	Burgundy Poly	50700
O	Coronado Gold Poly	22728
P	Silverglaze Poly	32603
Q	Verdoro Green Poly	43745
R	Regimental Red	71583
S	Champagne Poly	22813
T	Montego Cream	81578
Y	Mayfair Maize	81500

TWO-TONES: The first letter indicates the lower color, the second letter indicates the upper color.

1968 PONTIAC

A	Starlight Black	9300
C	Cameo Ivory	8259
D	Alpine Blue Poly	13512
E	Aegena Blue Poly	13513
F	Nordic Blue Poly	13514
G	April Gold Poly	22942
I	Autumn Bronze Poly	60517
K	Meridian Turquoise Poly	13517
L	Aleutian Blue Poly	13516
N	Flambeau Burgundy Poly	50775
P	Spring Mist Green Poly	43774
Q	Verdoro Green Poly	43745
R	Solar Red	71634
T	Primavera Beige	81617
V	Nightshade Green Poly	43773
Y	Mayfair Maize	81500

TWO-TONES: The first letter indicates the lower color, the secondletter indicates the upper color.

1969 PONTIAC

40	Mayfair Maize	81500
50	Cameo White	2058
51	Liberty Blue Poly	2075
52	Matador Red	2076
53	Warwick Blue Poly	2077
55	Crystal Turquoise Poly	2078
57	Midnight Green Poly	2079
59	Limelight Green Poly	2080
61	Expresso Brown Poly	2081
63	Champagne Poly	22813
65	Antique Gold Poly	2082
67	Burgundy Poly	50700
69	Palladium Silver Poly	2059
72	Carousel Red	2084
73	Verdoro Green Poly	2095
76	Goldenrod Yellow	2094
87	Windward Blue Poly	13759
10	Starlight Black	9300

GRAND PRIX COLORS

86	Claret Red Poly	71763
88	Nocturne Blue Poly	13841
89	Castillian Bronze Poly	23215

TWO-TONES: The first two digits indicate lower color, the next two digits upper color.

1970 PONTIAC

10	Polar White	8631

14	Palladium Silver Poly	2059
25	Bermuda Blue Poly	2165
26	Lucerne Blue Poly	2213
28	Atoll Blue Poly	2166
34	Mint Turquoise Poly	2168
43	Keylime Green Poly	2170
45	Palisade Green Poly	2171
47	Verdoro Green Poly	2095
48	Pepper Green Poly	2173
50	Sierra Yellow	2175
51	Golden Red Yellow	2094
53	Coronado Gold Poly	23211
55	Baja Gold Poly	2178
58	Granada Gold Poly	2179
60,(06)	Orbit Orange	2257
63	Palomino Copper Poly	2183
65	Carousel Red	2084
67	Castillian Bronze Poly	23215
75	Cardinal Red	2189
78	Burgundy Poly	50700
19	Starlight Black	9300

TWO-TONES: The first two digits indicate the lower color,the second two digits the upper color.

1971 PONTIAC

11	Cameo White	2058
13	Nordic Silver Poly	2327
16	Bluestone Gray Poly	2161
19	Starlight Black	9300
24	Adriatic Blue Poly	2328
26	Lucerne Blue Poly	2213
29	Regency Blue Poly	2330
39	Aquarius Green Poly	2331
42	Limekist Green Poly	2333
43	Tropical Lime Poly	2334
49	Laurentian Green Poly	2337
51	Golden Red Yellow	2094
53	Quezal Gold Poly	2339
55	Baja Gold Poly	2178
59	Aztec Gold Poly	2359
06, 60	Orbit Orange	2257
61	Sandalwood	2181
62	Canyon Copper Poly	2340
65	Carousel Red	2084
66	Bronzini Gold Poly	2367
67	Castillian Bronze Poly	23215
75	Cardinal Red	2189
78	Rosewood Poly	2350

TWO-TONES: The first two digits indicate the lower color, the second two digits indicate the upper color.

1972 PONTIAC

11	Cameo White	2058
14	Revere Silver Poly	2429
18	Antique Pewter Poly	2430
19	Starlight Black	9300
24	Adriatic Blue Poly	2328
26	Lucerne Blue Poly	2213
28	Cumberland Blue Poly	2166
36	Julep Green Poly	2433
43	Springfield Green Poly	2435
48	Wilderness Green Poly	2439
50	Brittany Beige	2441
53	Quezal Gold Poly	2339
54	Arizona Gold Poly	2442
55	Shadow Gold Poly	2443
56	Monarch Yellow	2444
57	Brasilia Gold Poly	2445
62	Spice Beige	2447
63	Anaconda Gold Poly	2448
65	Sundance Orange Poly	2450
69	Cinnamon Bronze Poly	2452
75	Cardinal Red	2189

TWO-TONES: The first two digits indicate the lower color, the second two digits indicate the upper color.

1973 PONTIAC

11	Cameo White	2058
19	Starlight Black	9300
24	Porcelain Blue Poly	2523
26	Regatta Blue Poly	2524
29	Admiralty Blue Poly	2526
42	Verdant Green Poly	2528
44	Slate Green Poly	2529
46	Golden Olive Poly	2530
48	Brewster Green	2531
51	Sunlight Yellow	2533
56	Desert Sand	2537
60	Valencia Gold Poly	2538
64	Ascot Silver Poly	2541
66	Burnished Amber Poly	2542
68	Burma Brown Poly	2543
74	Florentine Red Poly	2545
75	Buccaneer Red	2546
81	Mesa Tan	2549
97	Navajo Orange Poly	2555

TWO-TONES: The first two digits indicate lower color, the next two digits upper color.

1974 PONTIAC

11	Cameo White	2058
19	Starlight Black	9300
24	Porcelain Blue Poly	2523
26	Regatta Blue Poly	2524
29	Admiralty Blue Poly	2526
36	Gulfmist Aqua Poly	2640
40	Fernmist Green	2641
44	Lakemist Green	2642
46	Limefire Green Poly	2643
49	Pinemist Green Poly	2645
50	Carmel Beige	2646
51	Sunstorm Yellow	2677
53	Denver Gold Poly	2649
55	Colonial Gold	2650
59	Crestwood Brown Poly	2367
64	Ascot Silver Poly	2541
66	Fire Coral Bronze Poly	2653
69	Shadowmist Brown Poly	2656
74	Honduras Maroon Poly	2658
75	Buccaneer Red	2546

TWO-TONES: The first two digits indicate lower color, the next two digits the upper color.

1975 PONTIAC

11	Cameo White	2058
13	Sterling Silver Poly	2518
15	Graystone	2742
19	Starlight Black	9300
24	Artic Blue	2745
26	Bimini Blue Poly	2746
29	Stellar Blue Poly	2748
31	Gray Poly	2751
39	Burgundy Poly	2786
44	Lakemist Green	2642
45	Augusta Green Poly	2750
49	Alpine Green Poly	2752
50	Carmel Beige	2646
51	Sunstorm Yellow	2677
55	Sandstone	2755
58	Ginger Brown Poly	2757
59	Oxford Brown Poly	2758
63	Copper Mist Poly	2759
64	Persimmon Poly	2760
66	Fire Coral Bronze Poly	2653
72	Roman Red	2544
74	Honduras Maroon Poly	2658

75	Buccaneer Red	2546
80	Tampico Orange Poly	2548

In two-tone combinations the first two digits indicate lower color, the next two digits upper color.

1976 PONTIAC

11	Cameo White	2058
13	Sterling Silver Poly	2518
16	Medium Gray Poly	2862
19	Starlight Black	9300
28	Athena Blue Poly	2772
35	Polaris Blue Poly	2863
36	Firethorn Red Poly	2811
37	Cordovan Maroon Poly	2864
40	Lime Poly	2866
49	Dark Green Poly	2752
50	Bavarian Cream	2867
51	Golden Rod Yellow	2094
55	Anniversary Gold Poly	2861
57	Cream-Gold	2884
65	Light Buckskin	2829
67	Durango Bronze Poly	2871
72	Red	2544
78	Carousel Red	2084

TWO-TONES: The first two digits indicate lower color, the next two digits upper color.

1977 PONTIAC

11	Cameo White	2058
13	Sterling Silver Poly	2953
15	Gray Poly (Two-Tone)	2862
19	Starlight Black	9300
21	Lombard Blue	2815
22	Glacier Blue Poly	2955
29	Nautilus Blue Poly	2959
32	Royal Lime	2960
36	Firethorn Red Poly	2811
37	Cord. Maroon Poly (Two-Tone)	2864
38	Aquamarine Poly	2961
44	Bahia Green Poly	2964
48	Berkshire Green	2965

50	Cream Gold	2884
51	Goldenrod Yellow	2094
61	Mojave Tan	2869
63	Buckskin Poly	2970
64	Fiesta Orange	2968
69	Brentwood Brown Poly	2972
72	Roman Red	2973
75	Buccaneer Red	2546
78	Mandarin Orange	2976

TWO-TONES: The first digits indicate lower color, the second two digits upper color.

1978 PONTIAC

11	Cameo White	2058
15	Platinum Poly	3076
16	Gray Poly	3077
19	Starlight Black	9300
21	Dresden Blue	3078
22	Glacier Blue Poly	2955
24	Martinique Blue Poly	3079
29	Nautilus Blue Poly	2959
30	Lombard Blue (Skybird Only)	2815
44	Seafoam Green Poly	3081
45	Mayfair Green Poly	3082
48	Berkshire Green Poly	2965
50	Spec. Editional Gold Poly (Firebird)	3071
51	Sundance Yellow	3084
55	Gold Poly	82352
56	Burnished Gold Poly	3086
58	Blue	15109
61	Desert Sand	3088
63	Laredo Brown Poly	3090
67	Ember Mist Poly	3091
69	Chesterfield Brown Poly	3092
72	Roman Red	2973
75	Mayan Red	3095
77	Carmine Poly	3096
79	Claret Poly	3098

TWO-TONES: The first two digits indicate lower color, the next two digits upper color.

INDEX

If you have enjoyed this book, other titles you might find helpful with your painting or restoration projects include:

How to Paint Your Car, by David H. Jacobs, Jr.
How to Custom Paint: Techniques for the '90s,
   by David H. Jacobs, Jr.
Boyd Coddington's How to Paint Your Hot Rod,
   by Timothy Remus
How to Restore Your Collector Car, by Tom Brownell
How to Restore Your Musclecar, by Greg Donahue and
   Paul Zazarine
How to Restore Your Chevrolet Pickup, by Tom Brownell
How to Restore Your Ford Pickup, by Tom Brownell
Automotive Tools Handbook, by David H. Jacobs, Jr.
How to Repair and Restore Bodywork, by David H. Jacobs, Jr.
How to Restore Auto Upholstery, by John Martin Lee

Motorbooks International has dozens of books on painting and automotive restoration and repair. They are available at fine bookstores everywhere or from Motorbooks International. Call or write for a free catalog:

Motorbooks International
P.O. Box 1
Osceola, WI 54020
1-800-826-6600